CONNECTIONS and CONTENT

Reflections on Networks and the History of Cartography

Mark Monmonier

Esri Press
REDLANDS | CALIFORNIA

Esri Press, 380 New York Street, Redlands, California 92373-8100

23 22 21 20 19 1 2 3 4 5 6 7 8 9 10

Printed in the United States of America.

Library of Congress Control Number:2019945357

Ask for Esri Press titles at your local bookstore or order by calling 1-800-447-9778. You can also shop online at www.esri.com/esripress. Outside the United States, contact your local Esri distributor or shop online at eurospanbookstore.com/esri.

Esri Press titles are distributed to the trade by the following:

In North America:
Ingram Publisher Services
Toll-free telephone: 800-648-3104
Toll-free fax: 800-838-1149
E-mail: customerservice@ingrampublisherservices.com

In the United Kingdom, Europe, the Middle East and Africa, Asia, and Australia:
Eurospan Group
3 Henrietta Street
London WC2E 8LU
United Kingdom
Telephone 44(0) 1767 604972
Fax: 44(0) 1767 6016-40
E-mail:eurospan@turpin-distribution.com

Dedicated to the memory of my parents,
John Carroll Monmonier (1909–99) and
Martha Elizabeth Georgia Mason Monmonier (1912–89)

CONTENTS

PREFACE

For me, the notion of maps as networks, or interconnected systems, is deeply personal. Although my earliest experiences with cartography began with crayons and rolls of shelf paper on which I created dozens of long, thin farms—corn here, the barn there, and wheat over yonder—the leap from neophyte mapmaker to researcher/raconteur grew out of a fascination with the maps in train schedules and the discarded copies of the *Official Guide of the Railways* that Dad brought home from his office downtown at the Baltimore and Ohio Railroad. I was also intrigued by maps of Baltimore's streetcar and bus network—because we didn't have a car until we moved out to Baltimore County, where Mom taught school, public transit was what we used for Sunday afternoon outings. While we lived in the city, I became curious about how the fire alarm boxes on street corners were connected by a single wire running from box to box, which I mapped within our neighborhood and beyond, using my two-wheeler to discover multiple circuits leading to local firehouses. Years later, I learned that pulling an alarm would cause a spring-wound mechanism to send out a unique series of coded interruptions that pinpointed the particular box. Some small cities in upstate New York still use this mechanism to back up the 911 system.

In high school, I discovered US Geological Survey (USGS) topographic maps as a rich source of information on railroads, both extant and abandoned. In graduate school years later, I contemplated a master's thesis on railway abandonment but shelved the idea until I found time, several years after earning a PhD, to publish a short paper with the pompous title, "Railroad Abandonment in Delmarva: The Effect of Orientation on the Probability of Link Severance in a Transport

Network." My multipanel temporal map and related graphic model were early examples of what's now called a *story map*. And three decades ago, after I discovered that authoring books could be more fulfilling than writing journal articles, I drafted the first of several outlines for a book on maps and networks: a project that sat on the back burner until a couple of years ago, when the seven-chapter structure for *Connections and Content* came into focus. I'm glad I waited.

What made this latest plan compelling was a string of one-word chapter titles that relate technologies for observing and measuring the landscape to key principles for making or using maps. Chapter 1, which looks at how the scales of maps based on triangulation networks depend on precisely measured *baselines*, precedes the chapter on *geometry*, which examines mathematical connections between gravity, heavenly bodies, optical instruments, telegraph lines, and three-dimensional figures called *ellipsoids*, which outperform the sphere in describing our planet's shape. Chapters 3 and 4, largely focusing on rivers, canals, and railways, explore the trove of map *symbols* and the modes of map analysis whereby past and current transportation *infrastructure* was conceived, located, built, and advertised to the public. Chapter 5 examines the role of *telecommunications* in not only birthing and nurturing weather science but also shaping the institutional networks of operational forecasting. *Topology*, a mathematical genre dedicated to adjacency rather than distance, is a fitting title for chapter 6, which looks at cartographic databases designed to promote the collection and analysis of census data, the interactive display of terrain data, and the calculation of optimum routes for satnavs (in-vehicle satellite navigation systems). The final chapter, on *control*, explores the role of maps in describing the internet, enabling driverless cars and drones, preventing railway accidents, and designing congressional districts that let a minority of voters dictate results. Whoever promised that modern mapping is inherently benign?

Maps as networks is an ideal theme, or leitmotiv, for teaching and understanding the historical development of modern cartography, which evolved with the aid of mathematics and electronic computation from a clever integration of telescopes, rulers, and spirit levels to its present-day dependence on satellites, computers, databases, telecom links, and lidar (airborne radar that pings the surface below with thousands of laser beams). Networks that support measurement, navigation, surveillance, information management, and institutional operations account for mapping technology's broad impacts on politics, science, transportation, and warfare. It's useful to appreciate maps' humbler origins.

Because cartography has unavoidable mathematical underpinnings, some of the terms and concepts used here might seem alien, if not confusing. I've simplified and summarized where appropriate and tried to explain why the occasional rough passage is worth plowing through, or at least skating across if that works—scanning through the text (which all of us do) is more prudent than missing the stories that follow. Chapter 1, which introduces the book and concludes with an overview of the six other chapters, walks the not-so-thin line between readers turned off by mathematical details and purists annoyed by studied simplification. I titled it "Baselines," rather than "Introduction," because it's framed around the most carefully measured links in triangulation networks—the measurements used to establish the scale of the map. The astute reader will, I hope, appreciate the caveats about visual observation, measurement error, strength of network, systematic adjustment, and the like. Although mapmaking can be conceptually complex when high levels of precision and reliability are needed, all chapters incorporate numerous and comparatively digestible stories about people who helped cartography get to where it is.

Chapter 2, titled "Geometry," appropriately follows "Baselines" because it looks at how diverse networks and geometric objects, some two centuries old and others much more recent, provide the geographic coordinates essential for science, national defense, transportation, and commerce. Its narrative condenses numerous facts about how these networks and geometric figures work separately and in concert, about how mapping networks evolved as measurement technologies improved and user requirements exploded, and about how cartography intersects the fields of astronomy, mathematics, geophysics, and electronic communications. Although my condensed narrative might trouble readers eager for a more detailed treatment—*Cartography in the Twentieth Century*, the encyclopedia I edited as volume 6 of the massive History of Cartography series, runs to a million words—readers seeking a general appreciation of the role of networks in making and using maps will, I trust, appreciate how key themes introduced in chapter 2 underlie the installments that follow.

A word about the illustrations and endnotes: I devised some images to reinforce the verbal narrative and extracted others from historic sources to promote understanding and authenticity. Typically, I create or select images as I write, to integrate them closely with the text. Look at the illustrations carefully as you read: even the most intriguing facsimile excerpts from historic maps are intended as more than eye candy or window dressing. Because this book was originally planned to be printed in black and white, I occasionally adjusted the contrast to accentuate some features and deemphasize others, and sometimes I added an explanatory label.

A few other caveats: most of my examples are from the United States, which I know comparatively well and where most of my readers typically reside. I included a list of acronyms because they can be confusing despite an interpretation the first time they appear; having them all in one place, at the back of the book, seems helpful. But there is no glossary because concise parenthetical definitions, as with *lidar* and *satnav* in this section, seem sufficient. If a fuller explanation seems warranted, check the endnotes. Finally, because several readers of an earlier draft thought me needlessly apologetic for the occasional complex passage—mostly in the earlier chapters, thank goodness—I toned down the preface's mea culpa, mea culpa, mea *maxima* culpa (as we altar boys used to say). A bit of complexity is worth celebrating.

One reader thought my use of the term "network" might be confused with social networks. I hope not. A glance through the illustrations should quash that notion. Another reader mentioned a variety of cartographically relevant networks I've ignored or downplayed: networks such as industrial supply chains, urban transit systems, paths of warfare and the flight of refugees, airline hub-and-spoke connections, and the parcel boundary networks used for real-property assessment. A fair criticism, perhaps, but this book is not a textbook or manual but a personal reflection on networks historically important in the development of cartography—networks worth knowing to better understand the field's origins and perspective. I know from experience that general readers prefer shorter narratives, not massive treatises, and I've a few other projects on my literary bucket list to keep me busy.

Though I have probably forgotten numerous people who, over the years, have added to what I know about cartography and networks, a tally of "helpful people" that I started just before I began writing this book allows a reasonably complete list of acknowledgments. Librarians head my list of people due profound thanks: Albert "Skip" Theberge, at the National Oceanic and Atmospheric Administration's NOAA Central Library, in Silver Spring, Maryland; Michael Kline and Edward Redmond, in the Geography and Map Division, US Library of Congress; Katie Tomfohrde, at the Houston Public Library; Karl Longstreth and Tim Utter, at the University of Michigan; Susanne Caro, at the University of Montana's Mansfield Library; Bonnie Gallinger, in the William C. Wonders Map Collection, Larry Laliberte, the Data Spatial Services Librarian, and Elizabeth Wallace, a "Collection Strategies Librarian," all at the University of Alberta, in Canada; and closer to home, Darle Balfoort, Stephanie Helsher, Diane W. McKenney, John Olson, and Amanda Page, at Syracuse University's Bird Library.

Particularly helpful people at government agencies include Eric J. Butterbaugh, in the Public Affairs Office, Defense Advanced Research Projects Agency (DARPA); Jason Y. Kim, webmaster at the federal government's GPS Directorate; and Harold "Harry" Nelson, at the Maine Department of Transportation.

Richard Garland, a geodetic network and geocaching enthusiast in New York City, allowed me to reproduce his photograph of the North Base (Massachusetts) triangulation monument; and Jeff Otto of the Ontario and Western Railway Historical Society, in Middletown, New York, and Bill Dexter of the Oswego Railroad Museum, in Oswego, New York, were helpful in their due diligence to confirm that no one holds a copyright to the 1930s-era New York, Ontario, and Western Railway map.

In the geography department at Syracuse University, our staff cartographer, Joe Stoll, a consistently reliable source of advice on Adobe® Photoshop® and Adobe Illustrator®, improved several of the facsimile illustrations and drafted key parts of two maps in Adobe Illustrator; colleague Jane Read shared her knowledge of drones; Margie Johnson, our administrative assistant, ran interference with the accounting system; and the IT support unit in the Maxwell School, in particular, Brian von Knoblauch, Stan Ziemba, Mike Cavallaro, Daryl Olin, and Mike Fiorentino, helped me cope with our temperamental wireless network.

Old friends who contributed useful advice and subtle encouragement include Peter Collier, University of Portsmouth, UK; James Roger Fleming, Colby College; Matthew H. Edney, University of Southern Maine; Mary Pedley, University of Michigan; Susan Schulten, University of Denver; and Judith Tyner, California State University, Long Beach.

Many thanks to Esri® Press. Candace Lyle Hogan encouraged me to submit the manuscript to Esri Press and initiated the peer review before her retirement in January 2019. An insightful reading of the full manuscript by Patrick Kennelly, of Long Island University, is much appreciated, as are most of the suggestions of the other four, still anonymous reviewers. Publisher Catherine Ortiz and acquisitions supervisor Stacy Krieg clinched the deal, acquisitions editor Claudia Naber shepherded the project to editorial, and manuscript editor Carolyn Schatz provided observant catches and helpful suggestions, Frances Tann came up with an appealing cover design, and Monica McGregor designed the interior.

1

BASELINES

**What is a map? is a question that invites diverse answers. In *The Nature of* ***Maps*, a gospel in long-forgotten graduate seminars, Arthur Robinson and Barbara Petchenik sidestepped a needlessly narrow focus on Planet Earth or its physical environment by resurrecting an obscure French term for their definition, "graphic representation of the milieu."[1] No less basic is the triad "scale, projection, and symbolization," which framed more than a half century of map design courses.[2] But neither definition captures the flexibility and promise of late twentieth-century electronic maps designed for machines faster, if not more reliable, than the human eye/brain system.

A more encompassing definition must accommodate both the maps our eyes see and the maps a digital computer reads as data, as when software finds the shortest route between two points or crafts a politically advantageous yet legally acceptable configuration of voting districts. For the visual map's basic role in describing regions and connecting places, map symbols depict networks of nodes and links. For the electronic map's role in informing algorithms, these networks exist in computer memory as systematically organized data. Making a clear distinction between visual maps that depict networks and electronic maps that are networks of memory locations and digital addresses is more important than any compromise definition of map I could cobble together. Whatever its wording, that definition could not ignore the notion of networks.

That said, *network* has a much broader connotation in cartography because other kinds of networks provide an indispensable geometric framework for detailed topographic mapping or an integrated system for the efficient and timely

flow of geographic information from dispersed observation points to a widespread community of internet or wireless users. A classic example is the complementary radar and telecommunications networks that collect and distribute data for the weather map, which could not otherwise exist in any of its varied formats.

These diverse applications beg the question: What is a network? Although dictionaries suggest numerous answers, I like Supreme Court Justice Potter Stewart's famous retort when asked to define pornography: "I know it when I see it." Figure 1.1 illustrates three basic types of network, as represented cartographically.[3] The simplest is the *branching network*, exemplified by rivers and streams, which merge into generally wider channels as their waters flow from higher to lower elevations and from smaller to larger catchment areas. In network terminology, the points of confluence are *nodes* or *vertices*, and the intervening paths are *links*, *edges*, *chains*, *arcs*, or [stream] *segments*. By contrast, *circuit networks* have loops, like those that allow a motorist to circle the block or choose among several routes. There are also *barrier networks*, principally political or administrative boundaries, the segments of which block or constrain flows of goods, travelers, or migrants. Topographic maps, which integrate drainage, transportation, and political jurisdictions, have all three types, as do many web maps.

Some networks incorporate directional bias. Because water does not flow uphill, a stream network is a *directed* network, in contrast to a telecommunications or highway network, which is *nondirected* because two-way flow is the norm. But because of one-way streets, road networks can have both directed and nondirected links. And as pilots are aware, networks can be three-dimensional and include links that represent airline connections, which crisscross on a two-dimensional map without intersecting at nodes.

In contrast to stream and road networks, in which links can be highly sinuous, some networks connect their nodes with straight lines, as shown on the right side of figure 1.1, under "Circuit Network." On an abstract network map, sometimes called a *schematic diagram or planar graph*, the lengths of these straight-line links are not strictly proportional to the measurements on which the network is based; examples include road distance, travel time, and per-ton transport cost. Road maps and tour books often include small maps showing average driving time between principal cities. These cartographic caricatures help motorists compare different routes between widely separated travel points.

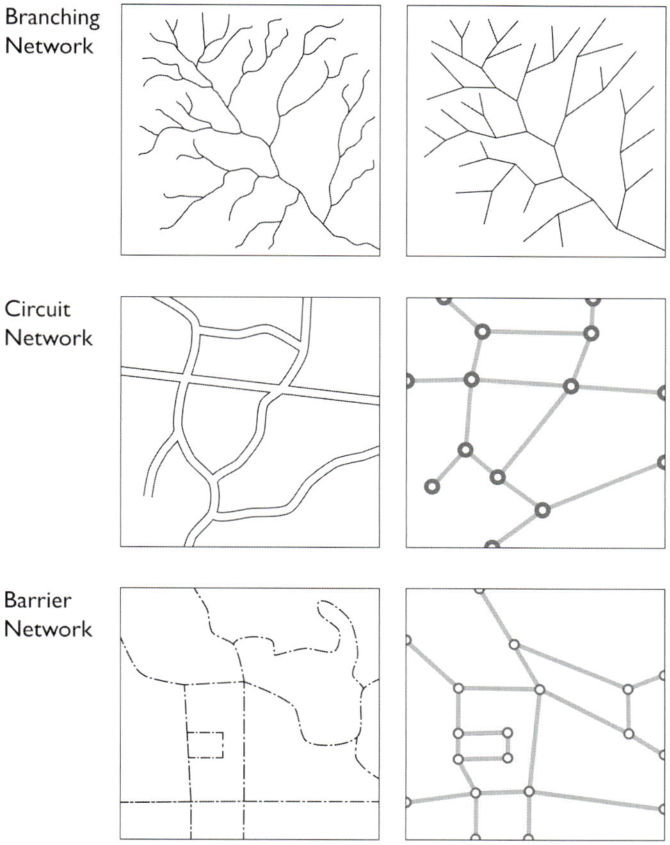

Figure 1.1. The three principal types of network, as portrayed on a map by geographically realistic links (*left*) or treated more abstractly as a planar graph (*right*).

Mapmakers go one step further by making some links directly proportional in length to the horizontal distance between nodes. As illustrated by the terrain profile in figure 1.2, the overland distance between nodes can be much greater than the corresponding horizontal distance, particularly in rough terrain. Cartographers call it *planimetric distance* because length is measured in a plane. Planimetric distance is the only distance reliably portrayed on topographic maps with a bar scale. The remainder of this chapter focuses on networks that frame planimetric maps.[4]

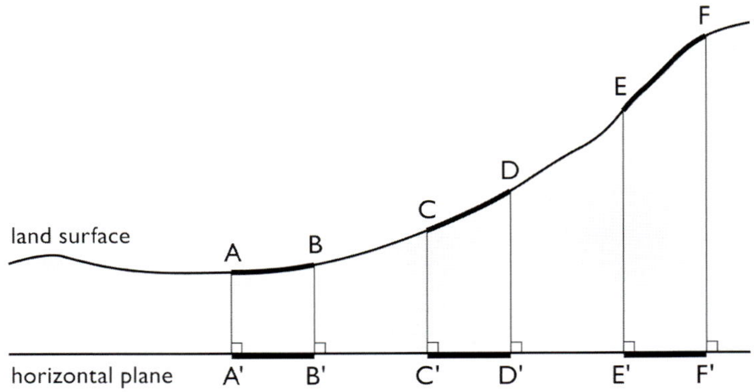

Figure 1.2. Points *A* through *F* on this hypothetical terrain profile project vertically onto a horizontal plane at points *A'* through *F'.* Overland distances *A–B, C–D,* and *E–F* are equal in planimetric distance despite longer overland distances resulting from progressively steeper slopes.

One strategy for making topographic maps employs an adjustable drawing board known as a *plane table,* shown in an 1865 engraving (figure 1.3) to which I added labels identifying key components, such as *clamps* for firmly anchoring a sheet of waterproof drawing paper. Mounted on a tripod, the drawing board can be rotated around a vertical line through the center of the tripod and made perfectly horizontal with a circular bull's-eye spirit level; when the bubble aligns with the circle on a small glass dome, the board is sufficiently level to represent the horizontal plane in figure 1.2. Some plane tables use a T-shaped spirit level, with a pair of linear bubbles fixed at right angles. Though no longer used, except perhaps in field method courses that lack newer technology, the plane table affords a straightforward visual explanation of key concepts important to an appreciation of cartography's historic roots in field observation of real-world landscapes treated as networks of surveying points connected by lines of sight.

The surveyor begins by drawing a line marking the direction from point *A,* the plane table's initial location, to a distant point *B* at which an assistant holds a vertical pole. Figure 1.4 describes the sequence of steps that follow. The surveyor sights through a device called an *alidade,* essentially a telescope joined to a straightedge aligned in the same direction. Once the pole is centered in the telescope's crosshairs, a line is drawn along the straightedge to mark the direction to point *B*. Point *A* is marked on the drawing, and the plane table is carried to point

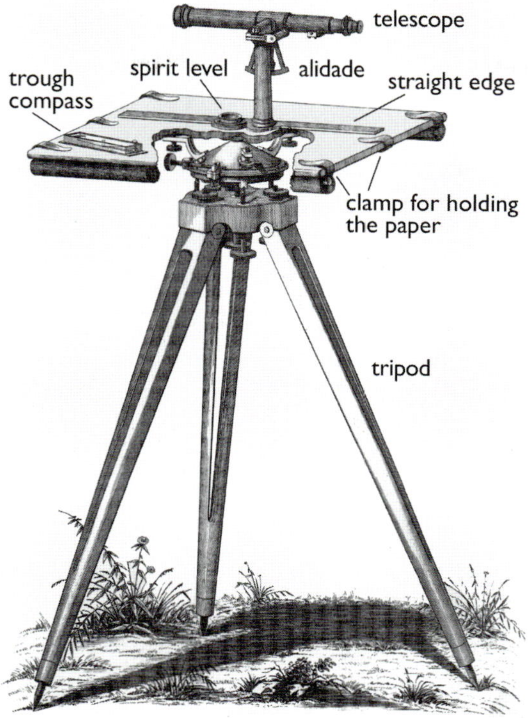

Figure 1.3. Plane table with a movable spirit level (on the drawing board, *toward the left*) and an alidade, with a straightedge for drawing a line aligned with its telescope.
NOAA Photo Library.

B and leveled. The surveyor then aligns the alidade with the original line drawn at *A* and rotates the drawing board until a pole set up at *A* is visible in the telescope. Marking point *B* along that line fixes the scale of the map.

Although the exact scale might not yet be known, all planimetric distances will be proportional to the length of line *A–B*. Determine that distance, or the distance between any pair of nodes, and you can calculate the map's scale as the ratio of map distance to ground distance, and then add a bar scale representing one or more typical distances in miles, feet, or kilometers. A carefully measured line used to establish the scale of the map is known as the *baseline*, highlighted as the title of this chapter. If the area mapped is not large, say only five miles east to west and a similar distance north to south, a conventional bar scale might show a mile divided into quarters, eighths, or tenths.

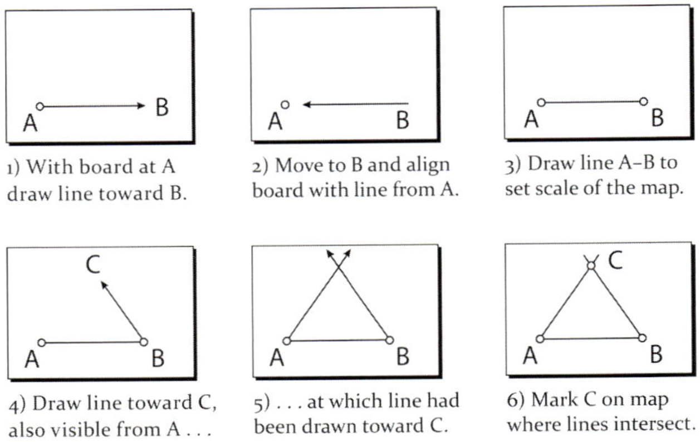

Figure 1.4. Steps in discerning the location of point *C* by drawing lines of sight on a plane table from points *A* and *B*.

The lower row in figure 1.4 shows how the surveyor adds other points to the map. At point *B*, he draws the line of sight to point *C*, toward which he had earlier drawn a line of sight from point *A*. Where the two lines intersect, he marks point *C* on the map. The plane table is a sixteenth-century innovation, rarely used in recent decades.[5] Nowadays, a surveyor uses a more complex but versatile instrument known as a *total station*. More about total stations later in this chapter.

William Gillespie (1816–68), a civil engineering professor at Union College, in Schenectady, New York, wrote a widely used nineteenth-century textbook on land surveying, in which he labeled this strategy, known as *intersection*, as "the most usual and most rapid method of using the Plane-table."[6] Gillespie used a simple line drawing (figure 1.5, *upper*) to show how intersecting lines of sight drawn on the plane table at two locations, *X* and *Y*, could fix the positions of additional points. On the lower right, a depiction of the drawing board at the second location, *Y*, shows a network with six nodes and twelve links. As Gillespie noted, another sheet could be placed on the drawing board and used to add points on the far side of this first line, *x–y*, to form a larger network.

Another diagram in his book (figure 1.5, *bottom*) describes how a second plane table method, called *progression*, could extend the network farther afield. If the length and direction of just one link, the baseline, was known with reasonable certainty, the enlarged network provided the skeletal framework for a multisheet map, which the surveyor could enrich by using lines of sight to sketch intervening

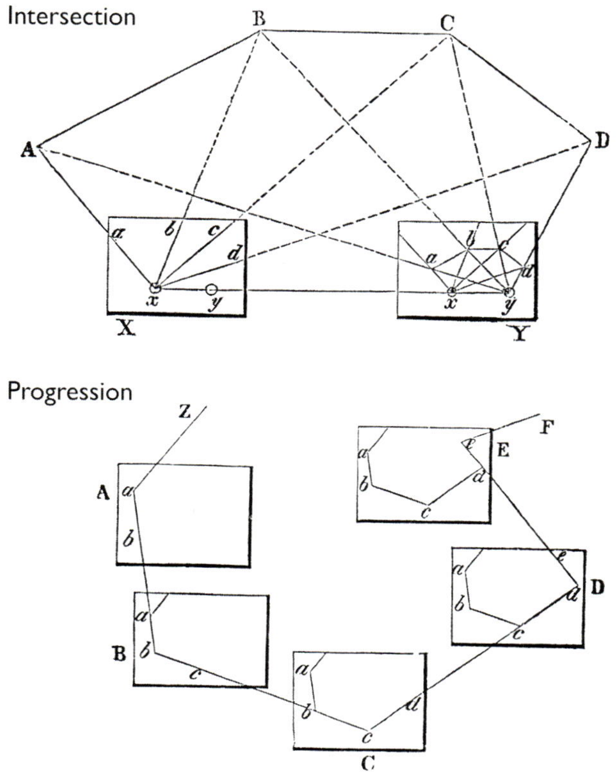

Figure 1.5. The plane table methods of intersection (*top*) and progression (*bottom*) as described in William Gillespie's 1855 textbook on land surveying. The *top* drawing illustrates how setting up the plane table in two locations can incorporate points beyond the initial line *x–y*. The *bottom* illustration shows how moving the plane table along a traverse can incorporate additional points along that traverse.

features. Indeed, a key advantage of the plane table is the ease with which a property surveyor or topographer can compose a map in the field, where it is easy to determine what's worth including. After completing his fieldwork, the mapmaker could carefully transfer all features onto a single sheet, typically drawn at a smaller scale. Reduction was straightforward if the original plane table drawings were on graph paper with four or five squares to the inch and the reduced drawing had a grid with twenty squares to the inch.

Plotting a composite drawing at a smaller scale diminished inaccuracies that arose in making the plane table plots, which were purely graphical solutions,

limited not only by the precision with which an alidade and sharp pencil can reproduce reliable lines of sight but also by the accuracy with which even a present-day surveyor can measure the length of a suitably long horizontal baseline. Although it is relatively easy to measure a short, essentially horizontal baseline, the resulting measurement is useful for a survey covering only a limited area—for example, a precise survey for reconstructing a crime scene or traffic accident. In mapping the site of a future strip mall, the surveyor might place a 25-foot measuring tape on level ground—if level ground can be found or made level by carefully joining and supporting a few long lengths of lumber. Even so, a 25-foot base is not useful for finding an intersection point (such as *C* in figure 1.4) much more than a few hundred feet away.

If you're uncomfortable with numbers, brace yourself: the story of cartographic networks cannot be told without getting into at least a few numbers. Field surveys involve observations that are also numerical measurements, which are an unavoidable source of error. Surveyors deal with error by averaging multiple readings and by measuring different parts of a network and checking for consistency, a process evident in everyday language whenever we assert the need to *triangulate* by looking at a problem from different viewpoints. Journalists call this process *fact checking*.

A surveyor's lengths and angles become useful only when leveraged by invoking basic trigonometry and simple statistics: concepts that might remind some readers of their first encounter with math anxiety. Recall trigonometry, that high school subject that ventured beyond plane geometry and basic algebra by obsessing over right triangles and introducing arcane terms such as *sine*, *cosine*, and *tangent*. Despite an undergraduate degree in mathematics, I am leery of confusing *sine* (opposite over hypotenuse) and *cosine* (adjacent over hypotenuse) because angular relationships are not part of a routine day. If you're from my generation, you might recall looking up these ratios in "trig tables." If you're my daughter's age and attended middle school in the mid-1990s, you might recall using a clunky multifunction calculator; nowadays, there's an app for it on your mobile device. Although trigonometric calculation was comparatively laborious for the nineteenth-century surveyors who established a national framework for detailed topographic mapping, their efforts depended on the fundamental mathematic concept of basic ratios shared by similar right triangles that vary in size.

No less basic is the notion that any non–right triangle can be partitioned into two right triangles with a common side. I'll spare you the formulas, but a bright ninth or tenth grader can prove that if you know two of the angles and one of the sides of a non–right triangle, you can calculate the lengths of the other two sides. Nowadays, a perhaps not-so-bright but nonetheless resourceful student will merely Google "irregular triangle calculator" to find a relevant app.[7] This notion is important because triangles in an actual survey network almost never include a right angle.

Trigonometry was useful in exploring my hypothetical 25-foot baseline's effect on the accuracy of locations estimated using intersecting lines of sight on a plane table. For this example, I imagined an intersection point 500 feet away from the endpoints of the 25-foot baseline. The resulting isosceles triangle has an acute angle measuring a mere 2.9°—I am rounding here to one decimal place—and angles of 88.6° at the ends of the baseline.[8] If the lines of sight are perfect, the intersection would lie 4,999 feet from the center of the baseline. But if the two lines of sight drawn on the plane table underestimate their angles by only a tenth of a degree (88.5°, instead of 88.6°), the distance from the intersection to the baseline shrinks to 467.2 ft.: an error of 32.6 ft.! Increasing the baseline to 100 ft. lowers the tenth-of-a-degree error to 8.6 ft., and increasing it to 500 ft., thereby forming an equilateral triangle, would further reduce the error to only 1.7 ft.

Surveyors are understandably concerned about what Gillespie called "ill-conditioned" triangles, in which an angle less than 30° or more than 120° makes graphical or numerical estimates highly sensitive to minor inaccuracies.[9] In an ideal, well-conditioned network, almost all triangles are nearly equilateral.

Surveyors further minimize error by measuring angles with a *theodolite* or *transit* and using numerical measurements and trigonometry to calculate the lengths of links and locations of nodes in a triangulation network.[10] Despite little agreement on the difference between a theodolite and a transit, the former term generally refers to a more accurate instrument. Nowadays, theodolites and transits are electronic devices designed to display and record digital measurements, but older devices measured horizontal and vertical angles by relating a telescope's line of sight to horizontal and vertical circles divided into degrees by evenly separated marks inscribed on glass or metal. In general, the greater the diameters of these horizontal and vertical circles, the more accurate the readings. But before any

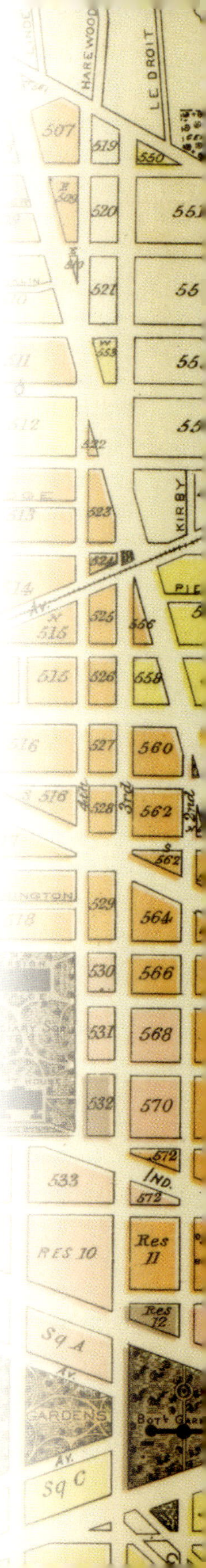

readings were taken, the instrument had to be carefully leveled, typically using three or four leveling screws at the top of the tripod and perpendicular spirit levels, as shown in figure 1.6, a comparatively Spartan mid-nineteenth-century theodolite used principally for measuring horizontal angles. More versatile instruments included magnifying lenses for reading the horizontal and vertical circles, which could be read to "single minutes or even less," according to Gillespie.[11]

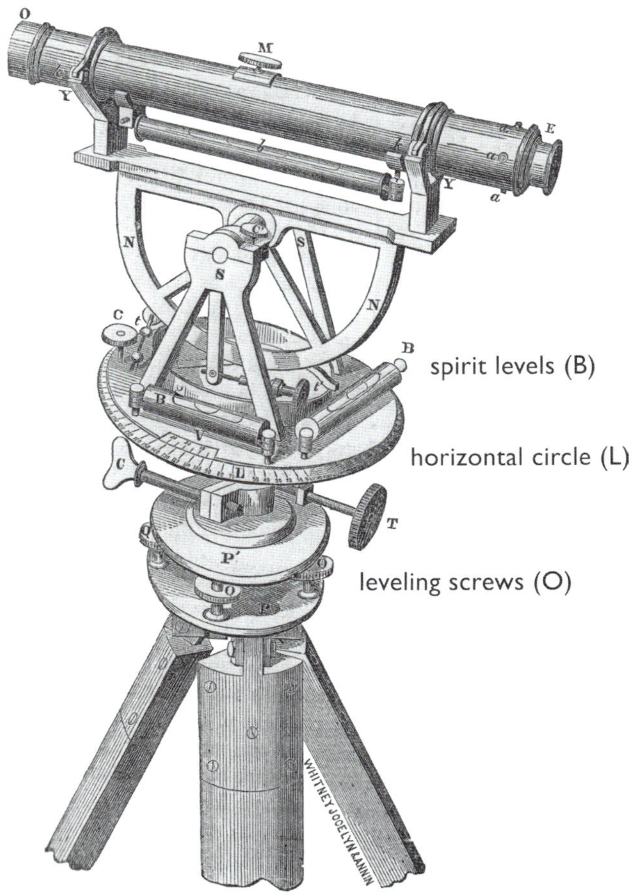

Figure 1.6. Engraving of an "English form" theodolite in William Gillespie's 1855 surveying text. Labels (*added on the right*) highlight the otherwise obscure locations of the leveling screws (*O*), spirit levels (*B*), and horizontal circle (*L*). Present-day theodolites typically use only three leveling screws.

Because an arc minute is a mere one-sixtieth of a degree, theodolites provide reliable observations for networks with much larger triangles than appropriate for a plane table. Consider the example, a baseline 5 mi. (26,400 ft.) long serving as one side of an equilateral triangle. If the angles at opposite ends are measured flawlessly, trigonometry calculates a distance of 22,863.1 ft. from the center of the baseline to the observed vertex. Underestimating the angles at the ends by one-sixtieth of a degree (59.983° instead of 60°) shortens the distance by only 15.4 ft. Increasing the side of the equilateral triangle to 10 mi. (52,800 ft.) merely doubles the discrepancy. For many purposes, the resulting error of 30.7 ft. would be insignificant. For example, if the map was to have a scale of 1:50,000 (1 in. on the map representing 50,000 in. on the ground), the center of a point symbol would be off by a miniscule 0.007 in.: less than a fifth of a millimeter and far, far smaller than the symbol plotted. Although an error of 31 ft. is potentially relevant to a surveyor who intends to install a boundary marker, the monument's exact location would need to be verified by observations from other points in a denser triangulation network.

To illustrate how trigonometric triangulation works, Gillespie included an excerpt from a US Coast Survey map describing the primary triangulation network for New England (figure 1.7). To place this configuration of surveying points and lines of sight in a geographic context, note the label "Atlantic Ocean," which places east toward the bottom, and the progression Rhode Island–Massachusetts–New Hampshire, which puts north toward the right. Gillespie's diagram describes the middle part of a massive project organized as a hierarchy of networks, with the more precise, primary triangulation connecting three baselines: one on Fire Island, off the south coast of Long Island, New York; another, along a relatively level 10¾-mile stretch of the Boston and Providence Railroad in Massachusetts; and a third, over unavoidably more rugged terrain at Epping Plain, Maine, near the Canadian border. Links connected local high points visible from each other over distances as long as fifty miles—visible at night, that is, from a portable tower that elevated the theodolite above trees and other obstacles. In a triangulation network, each node must be visible from at least two other nodes. Gillespie's diagram is important because it describes a primary, or first-order, survey that framed secondary and tertiary surveys, which in turn anchored more detailed mapping with a plane table.

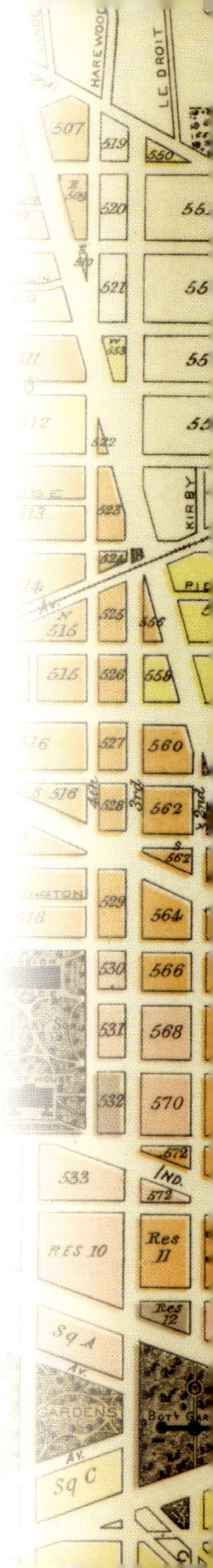

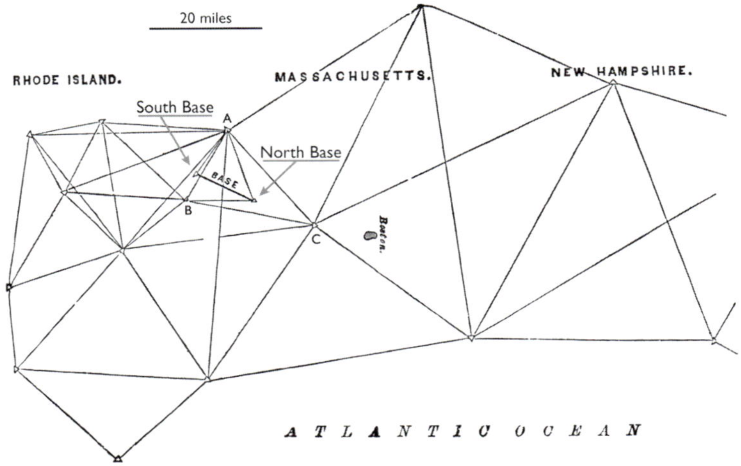

Figure 1.7. Southern portion of the US Coast Survey's primary triangulation for Section 1, extending from eastern Maine to Rhode Island, as shown in in William Gillespie's 1855 textbook on land surveying. North lies roughly *to the right*. The bar scale, labels identifying North Base and South Base, and markings for stations *A*, *B*, and *C*, mentioned in the discussion, were added by the author.

For each observation, the theodolite's telescope was focused on a signal mast positioned at a stationary location, or *station*, at which a triangulation tower was erected. Locations were selected after a reconnaissance survey verified that no intervening hill or structure would block the line of sight along a link. Each node's exact location was marked by a permanent monument not likely to disintegrate in bad weather or be easily carried off. The Coast Survey's annual report for 1854 described a typical monument as consisting of "an underground brick structure, enclosing a freestone block, in which a copper bolt is sunk [with] the extremity of the base .... marked by crosslines on the bolt head, over which a stone cap is laid."[12] Other designs were similarly immovable. If intended for daytime observation, the signal mast had to be sufficiently tall to be visible in the telescope and stand out against its background. Gillespie described a signal intended for nighttime observation that directed oxygen gas onto an alcohol flame amplified by a parabolic mirror and visible 66 miles away.[13]

In organizing its triangulation survey of the East and Gulf Coasts, the Coast Survey established nine project areas called *sections*, with Section 1 covering the coasts of Maine, New Hampshire, Massachusetts, and Rhode Island. Gillespie's drawing was based on the southern portion of the primary network for Section 1 as it existed in 1854.[14] (Other stations and links were added over the following decade.) Most of the stations had names such as Indian Hill (on Martha's Vineyard) and Quaker Hill (near Newport, Rhode Island), abbreviated Indian H. and Quaker H., with the abridgement *H* confirming that hills were natural places to erect triangulation towers. By contrast, the baseline's endpoints, chosen because they bracketed a long, straight, and generally level railway grade, were named North Base and South Base. In figure 1.7, North Base is on the right of South Base and closer to the bottom of the page, which seems counterintuitive because north is more commonly at the top of the map. In clipping a representative segment out of the larger Section 1 network, Gillespie rotated the map so that north lies to the right and south to the left. Because adjoining portions of the larger network had to fit together, the Fire Island baseline was in Section 2, which extended from Connecticut to Delaware.

Baselines are necessarily shorter than most other links in a triangulation network. Figure 1.7 shows how smaller triangles and trigonometric calculation helped integrate the Massachusetts baseline into New England's primary network. Careful examination of the upper-left portion of the figure shows how the triangle that includes the baseline was used (along with its measured angles) to calculate the length (13.13 mi.) of the side opposite the angle at South Base. Try it yourself.[15]

Although the side opposite North Base is not part of any other triangle, the length opposite South Base could be used to calculate distance *A–B* (14.43 mi.), which was then used in calculating distances *C–A* (22.85 mi.) and *C–B* (23.58 mi.), used in turn to calculate lengths for adjoining triangles and so on. Reliability at each step depends not only on accurately observed angles for the new triangle but also on the distance calculated in the last step. Whatever error occurs in measuring the baseline will be magnified as this basic length is propagated outward, step by step, through successive calculations. A small amount of uncertainty is inevitable, but mapmakers understand the importance of a highly reliable original length.

A note on the map that Gillespie had copied explained that many of the triangles were also parts of *quadrilaterals* (four-sided polygons), "both diagonals of which are determined," that is, treated as links in the triangulation network.[16]

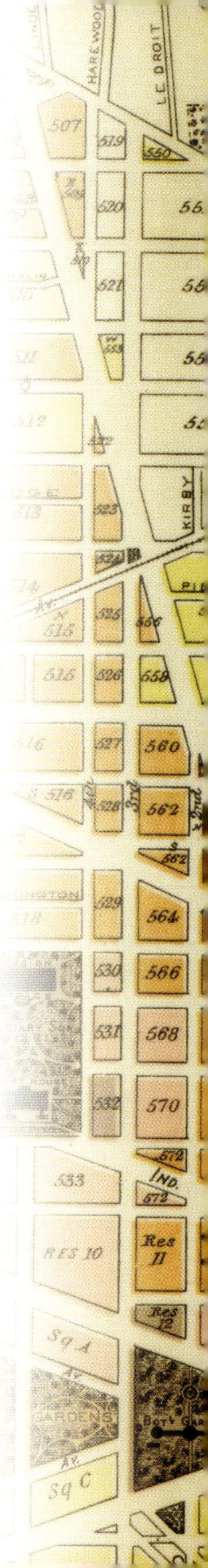

This strategy "strengthened" the network by affording additional checks on measured angles and calculated distances. For example, in much the same way that the angles of a plane triangle must sum to exactly 180°, the interior angles of a quadrilateral should add up to 360°. The additional links also helped Coast Survey scientists check calculations carried outward from baselines in different parts of the network using trigonometry and measured angles. Multiple baselines were a further source of strength, insofar as baselines at opposite ends of the Section 1 network, as well as near the center, minimized the step-wise propagation of length errors from one triangle to the next. The Coast Survey's annual report for 1865, in announcing the completion of primary triangulation for Section 1, applauded its three baselines as "peculiarly independent of each other" and valuable because "the estimates of the accuracy of the results .... may be received with confidence."[17]

Measuring the Massachusetts baseline took three months in the fall of 1844.[18] The project was directed by Edmund Blunt (1799–1866), a civilian "assistant" in what was then called the Survey of the Coast. (Although his job title might suggest a low-level lackey, the typical assistant was skilled in mathematics and measurement, and survey parties included more junior staff called '*aids*' or *subassistants*.) Blunt visited the Seekonk Plain, an area west of Boston near the border with Rhode Island, which an earlier, uncompleted survey for the state of Massachusetts had deemed suitable for a straight baseline of approximately five miles. Although a reconnaissance of the proposed baseline revealed irregular ground and woodland that would complicate the work, Blunt serendipitously discovered a straight stretch of double-track railroad still under construction. The timing was perfect because only one track had been laid along the recently graded right-of-way, and the railway's president and its superintendent willingly cooperated. In addition to expediting the work, the railroad afforded an opportunity to measure a baseline almost eleven miles long, a distance not surpassed for several decades.[19] Blunt selected endpoints, North Base and South Base, that could be connected by lines of sight to the larger triangulation network.

Measuring a baseline was a painstaking process of advancing a cumulative measurement forward in increments using precision metal bars that were carefully positioned end to end and held perfectly level and in a perfectly straight line using movable tripods called *trestles*. Figure 1.8, an etching from the historic imagery collection of the National Oceanic and Atmospheric Administration (NOAA), describes the process for the Epping baseline, surveyed in 1856 and 1857 at the eastern end of the New England network, at Epping Plain, Maine. Note the two measuring tubes, each containing a metal bar six meters long, and the six trestles, designed to sit on metal plates. The two trestles supporting a tube could be adjusted upward or downward as needed to keep the line level. Instruments in a box at the end of the tube assured the precise making and breaking of contact as the tubes leapfrogged along a carefully cleared path. Note the two trestles positioned ahead of the forward tube, ready to receive the aft tube, which was carried forward after breaking contact. *Leapfrogging* is an apt metaphor.

Figure 1.8. Coast Survey party measuring the northeastern baseline at Epping Plain, Maine. Leveling tripods called *trestles* supported canvas-covered tubes containing measuring bars. NOAA Photo Library.

An 1868 Coast Survey report described a carefully choreographed process involving ten men, a horse, and various pieces of equipment.[20]

> One assistant, to make the contact, give the signals, etc.
>
> One aid (sic), to align the bars, using a transit.
>
> One aid (sic), to record the inclination, temperature, and number of the bar, and, when measurement stops or halts for the day, to transfer the end of the rod to copper tack in stub, employing, for this purpose, the other transit.
>
> Two men, to carry the bar.
>
> Two men, to pick up the trestles, carry them forward, adjust them in line, and level them.

> One man, to attend the aid in charge of alignment, bring up [the] instrument, etc.
>
> One man, to keep up the transfer transit, and to be provided with stub, axe, and copper tack for an emergency, and to assist generally.
>
> Cart, horse, and driver, for the transportation of [the] heavy wooden box, in which the bars are kept when not in use; of water, stubs, spades, and tools, and of tent, in case of sudden storm.

The report's author, Coast Survey scientist Richard D. Cutts (1817–83), had been a brevet brigadier general during the Civil War and, no doubt, appreciated precision teamwork.

For the Massachusetts baseline, Blunt was the last to use an older measuring apparatus based on four bars two meters long and designed by Ferdinand Rudolph Hassler (1770–1843), who had founded the Survey of the Coast and directed it until his death in 1843. A Swiss surveyor and mathematics teacher who immigrated to the United States in 1805, Hassler insisted on measurements in meters, not feet. The Coast Survey's report for 1865 mentioned a combined length of 7.99987165 m at 32° F, calibrated against an 84-in. brass scale manufactured in London for the Survey of the Coast and brought to Washington in 1815.[21] Temperature readings from eight thermometers, two attached to each bar, were used to collect data so that the surveyors could adjust for thermal expansion, which made the bars a bit longer. The bars were placed end to end in a box 8 m long, and an elaborate set of tripods and levels was used to count the number of bars added incrementally as the apparatus advanced forward.[22]

Distilling multiple observations into a single measurement was an elaborate process. An initial estimate of 17,320 m, calculated by multiplying 8 m by 2,165 (the number of carefully counted incremental advances), required several corrections and adjustments.[23] Subtracting the "defect" of 0.00012835 (the amount by which the combined bar fell just short of 8 m) reduced the total by 0.2779 m, but because of an average temperature of 58.85° while the line was advancing, 3.2383 m was added to reflect thermal expansion. Because the railway grade was not precisely level, 0.5629 m was subtracted to yield the actual planimetric distance. And because the baseline was a bit longer than 2,165 increments, 3.9999 m (the length of two 2 m bars) was added between the end of the last increment and the North Base marker, and 0.0003 m was deducted to correct for "10° of temperature below 32°." Finally, an "additional scale measure" of 0.1012 m was added because the length was slightly longer than two 2 m bars, and 0.1220 m was subtracted to reduce the distance from an average elevation of 44.83 m to the "half-tide level of the ocean." Because of Earth curvature, a distance measured above sea level was a wee bit longer than the distance at sea level on a spherical planet. Coast Survey scientists tried to account for all possible errors.

Like true scientists, they then appended an error term based on the "probable error" of the four 2 m bars, for rising and falling temperatures while advancing the bars, and the probable error of the microscopes used to confirm a precise contact between leapfrogging bars. The result was an official measurement of 17,326.3763 ± 0.0358 m.

Nineteenth-century mapmakers responsible for the reliable depiction of the nation's coast also understood that efficient use of limited personnel and equipment required a hierarchy of networks, with a more carefully surveyed primary network anchoring a denser but less rigorously measured secondary triangulation that supported an even denser, not quite so intensively observed, tertiary network. The third network, in turn, framed the topographic mapping of land above the shoreline as well as the hydrographic charting of channel depth and submerged hazards. This hierarchy of systematically designed networks, each with a specific standard for reliability, differentiates geodetic surveying from comparatively ad hoc land surveys focused on property lines, roads, railways, canals, and bridges.

Cutts, who had described the teamwork required to advance the bars, also disclosed marked differences in triangle size among the primary, secondary, and tertiary networks.[24] In the primary network, the sides of triangles were between 20 and 146 mi. in length, in contrast to a range of 5 to 40 mi. for the secondary network and lengths generally shorter than 6 mi. for the tertiary network.

Cutts emphasized that triangulation was not the only responsibility of the geodetic survey. Highly reliable astronomical measurements of latitude and longitude, carried out at selected primary stations, were propagated mathematically to other parts of the network so that quadrangle maps could be properly bounded by specific parallels and meridians. Moreover, "from this special class of the geodetic work, subject to the least probable error, the dimensions and figure of the earth are deduced, as in the measurements, made and in progress, of the arcs of the meridian in the eastern, middle, and southern states, and of the Arc of the 39th Parallel across the continent." (In geodetic surveying, an arc is a chain of triangles

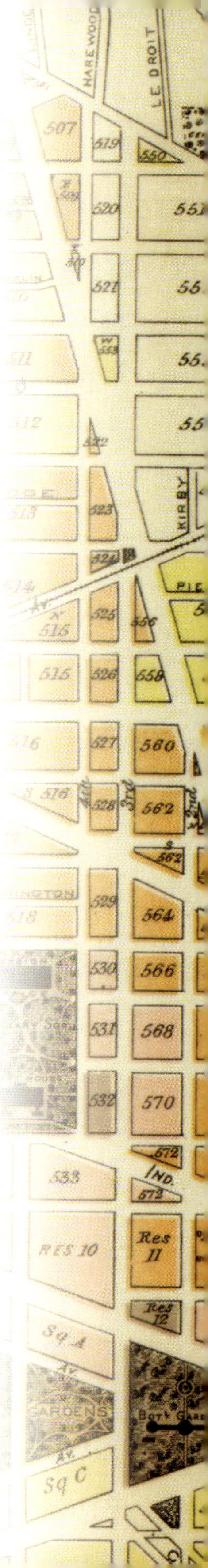

that follows a meridian, parallel, or other curved line that it is intended to measure.)[25] With responsibilities no longer confined to coastal states, the agency was renamed the US Coast and Geodetic Survey in 1878.

Cutts also described the cleverly designed portable towers that expedited the accurate measurement of angles at the network's nodes.[26] A tripod that supported the theodolite was surrounded by scaffolding that supported the observer, and accuracy required that tripod and scaffolding not touch. Towers as tall as 60 ft. not only elevated the line of sight above nearby terrain and trees but also compensated for the disappearance of distant objects beneath the horizon, as occurs at sea. Though a properly assembled tower was appropriately sturdy, I doubt that I could overcome my fear of heights sufficiently to climb one without considerable coaxing and shaming.

Because of Earth curvature, the relatively large triangles of the primary and secondary networks must be treated as spherical triangles, for which the sum of their angles is greater than 180°. This effect, known as *spherical excess*, is greater for larger triangles and can be predicted from their area as well as their latitude, which is relevant because the planet is flattened slightly toward the poles—think of a mandarin orange. For spherical triangles identical in area, the spherical excess will be measurably smaller at higher latitudes, and a precise geodetic survey must adjust for latitude.

To diminish instrumentation and human errors, the surveyor took multiple readings, half as "direct" readings, with the theodolite's telescope in the normal position, aimed toward the distant station, and half as "reverse" readings, with the telescope rotated 180° around the vertical circle and then 180° around the horizontal circle so that it was still sighted on the same distant station. This trick forces some instrumentation errors to work against each other and, in large measure, cancel each other out, but the geometry involved in designing, setting up, and reading a theodolite is complex, and I have necessarily simplified it.[27] Taking multiple readings, half in direct mode and half in reverse mode, is important because subsequent readings of the same angle are likely to be ever so slightly different—a single reading is just an observation, but multiple observations, when averaged, become a measurement. What's more, the variation of these readings about their average, or mean, is a useful estimate of the error incurred in making the measurement.

Geodesists make two kinds of adjustments. More straightforward is the *station adjustment*, which satisfies the basic rules of geometry by making certain that all angles surrounding a node sum to 360° and that the three angles in a plane triangle sum to 180°. More complex is the *figure adjustment*, which must accommodate not only the individual node and its adjoining triangles but also their neighboring nodes and triangles, all subject to errors of measurement. According to University of Pennsylvania professor Edward Lovering Ingram (1862–1938), the necessary adjustment "could be made in an infinite number of ways, but the adjustment that is sought is the one that assigns the most probable values to the various angles in view of their actually measured values."[28]

Ingram outlined the strategy in his 1911 textbook, *Geodetic Surveying and the Adjustment of Observations (Method of Least Squares)*. The process begins by removing (or apportioning) the spherical excess, so that subsequent calculations can be based on plane triangles. It then corrects for likely errors of measurement, based on each angle's weight, or "relative worth," calculated from the number of observations—the more observations the better. Ingram's computational procedure, known as *least squares*, recognizes that a perfect fit is unlikely: for any hypothetical trial adjustment of the network, each angle in the adjusted network will differ (ever so slightly, in most cases) from its observed value (after spherical excess is removed). For a satisfactory adjustment, most of these differences, known as *residuals*, will be small, but a few might be relatively large and potentially troubling. The method of least squares addresses this problem by finding a solution that minimizes the sum of the squares of the residuals.[29]

Anyone who has taken a course in elementary statistics is familiar with least squares, widely used to fit a straight line to points in an x,y graph. In a typical example, the horizontal coordinate might represent a person's years of formal education while the vertical coordinate reflects annual income, and the points describe a random sample of adults between 50 and 60 years old. Because the relationship between education and income is not perfect, a straight line representing the overall trend might not pass directly through any of the points. Although we could try using a straightedge to fit a line ourselves, least squares yields a "best fit" line that satisfies most researchers, who appreciate a standardized, presumably unbiased solution. Geodesists are similarly satisfied with the least-squares adjustment of triangulation networks.

How good was the Massachusetts Base and the New England network? Longtime Coast and Geodetic Survey mathematician Charles Anthony Schott (1826–1901) addressed this question in a report on the Eastern Oblique Arc of the United States. An endeavor that took two-thirds of a century to complete, this massive network stretched from the Gulf of Mexico near New Orleans to the Canadian border in eastern Maine (figure 1.9).[30] In a table summarizing the reliability of the arc's 483 triangles, Schott listed each section's average error of closure, a concept familiar to land surveyors. In a property survey, the series of line segments that enclose a parcel can be plotted sequentially, one after the other, starting at and returning to the traverse's point of beginning. Because measurement error makes precise closure highly unlikely, the distance from the ending point of the last boundary segment to the starting point of the first segment is called the "closing error," and is usually distributed throughout the traverse.[31] In a network of triangles, the closing error of any one triangle is the net sum of the corrections (some positive, some negative) made during figure adjustment, which can be averaged for all triangles in a regional network.[32] Schott went a step further by calculating the "probable error of an observed direction," which he considered "a more precise measure of accuracy."[33] His table showed the New England section, encompassing 53 triangles linking the Epping, Fire Island, and Massachusetts baselines, with a probable error of only ±0.26", matched only by the Dauphin Island base net, a small section in coastal Alabama with only five triangles—although the double-prime (") symbol is commonly associated with inches, in geodesy it refers to arc seconds. For comparison, the weighted mean for the entire arc is ±0.51", no less impressive insofar as one arc second is a mere 1/3,600 of a degree.

A folded map in Schott's 1902 report portrayed the arc as a long, thin polygon bounded by the sides of triangles along its outer edge (figure 1.9). The cartographer focused on the network's extent and coverage by omitting interior nodes and links and by filling the polygon with a blue tint, except for areas near the network's six baselines, which are tinted orange. Because the original map is over 18 in. from left to right—making it impossible to preserve a moderately detailed coastline and the names of state capitals in a significantly scaled-down rendering—I redrew its essential features, in black and white, to fit this book. These include the 1,623-mi. "geodetic line" along a great circle (shortest-path route)

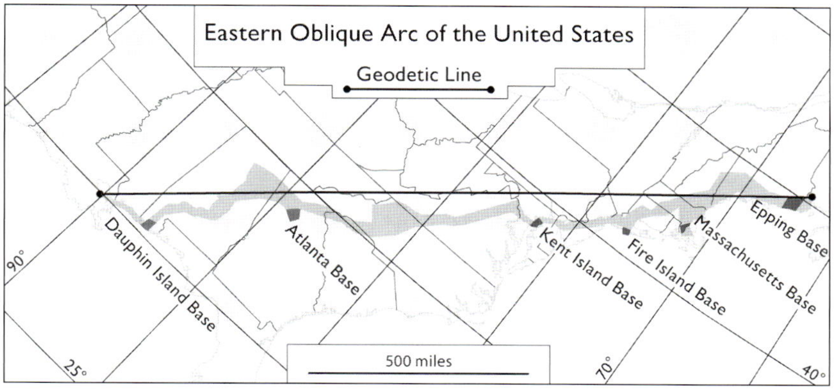

Figure 1.9. The Eastern Oblique Arc, redrawn at a much reduced size from a 1:7,000,000 map in Charles Schott's 1902 report. The darker shaded area encompasses the arc's 483 triangles.

from Calais, Maine, to New Orleans, at opposite ends of the arc, and meridians and parallels 5° apart, which show that north is toward the upper right and west is toward the upper left. In places, the arc is no wider than about 30 mi., and the longest stretch between baselines runs along the eastern edge of the Appalachians into northern Georgia. Note the westward bulges to incorporate Mount Washington (in New Hampshire), Mount Mitchell (in North Carolina), and several high-elevation stations in northern Alabama. Particularly prominent is the southeastward bulge to include the Massachusetts Base, initiated before the Coast Survey decided to integrate numerous regional networks into a national network that was also useful in calculating a more precise description of the earth's shape, known to resemble a sphere flattened slightly at the poles. Although triangulation arcs typically run hundreds of miles along a parallel or meridian, an oblique arc is especially useful in calculating the size of a three-dimensional figure formed by rotating an ellipse about its shorter axis.[34]

Aside from labeling the arc's six baselines, Schott did not explain the orange areas on his map (dark gray on mine). In figure 1.10, an enlarged facsimile excerpt focused on eastern Massachusetts, I used the grid of parallels and meridians to add the approximate location of the Massachusetts baseline, situated in the western part of its orange area.[35] A small map in Schott's report, titled *Massachusetts Base, and [Its] Connection with the Primary Triangulation of Massachusetts*, suggests

that this patch of orange contains the triangles in the primary network immediately adjacent to the baseline.[36] Although similar maps for the other baselines confirm this interpretation, Schott apparently wanted to highlight the locations of baselines too short to show up without enhancement on his 1:7,000,000 map.[37] This interpretation is consistent with the continuation of the blue (lighter gray in figure 1.10) part of the arc on the eastern side of the orange area, where the primary triangulation ran seaward to connect secondary triangulations with the main network.

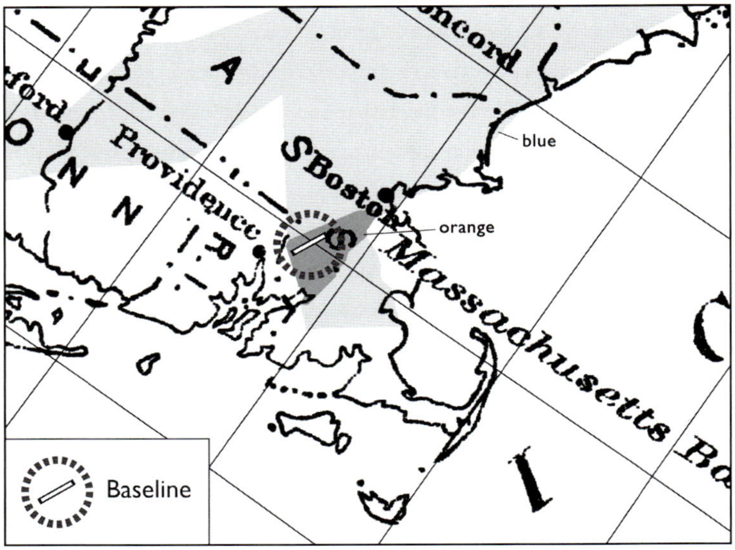

Figure 1.10. Enlarged excerpt from Schott's 1902 map in the vicinity of the Massachusetts baseline. Parallels and meridians were reconstructed and the baseline symbol was added by the author. The dark-gray area, colored orange on Schott's map, apparently encompasses the triangles immediately surrounding the baseline.

Maps in a 1922 report by Coast and Geodetic Survey mathematician Orlando Platt Sutherland (1889–1967) revealed impressive enhancements to the Massachusetts triangulation, which had grown significantly denser since the mid-nineteenth century.[38] One map described a "precise network" with thirty-nine links connecting nineteen nodes, including two in New Hampshire and one each in New York, Rhode Island, and Vermont. Sixteen additional maps covering smaller parts of the state showed hundreds of primary and secondary triangles linked to the more accurate network. Figure 1.11, a much reduced facsimile of the map that includes the Massachusetts baseline, reflects significant regional variation in network density, partly explained by spatial variation in prominent targets, such as church spires, tall chimneys, water towers, and other high points that can be triangulated from multiple stations and need not be occupied with a theodolite. Once the location of one of these secondary points was calculated and published, it could be used to triangulate other locations relevant to topographic mapping and construction surveys.

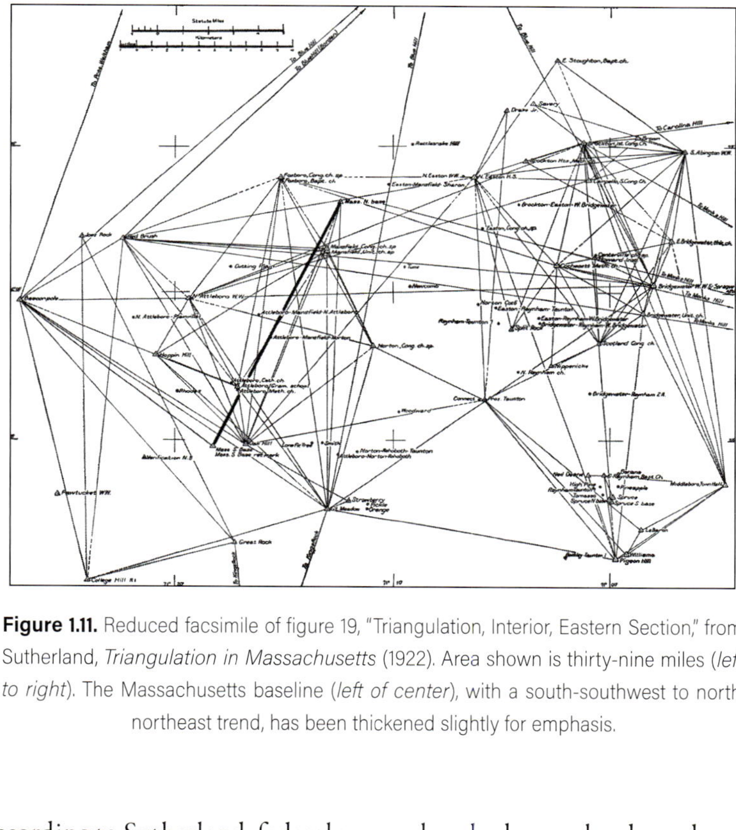

Figure 1.11. Reduced facsimile of figure 19, "Triangulation, Interior, Eastern Section," from Sutherland, *Triangulation in Massachusetts* (1922). Area shown is thirty-nine miles (*left to right*). The Massachusetts baseline (*left of center*), with a south-southwest to north-northeast trend, has been thickened slightly for emphasis.

According to Sutherland, federal mapmakers had recently adopted a new hierarchy of accuracy levels. Instead of the primary, secondary, and tertiary networks described by Richard Cutts in 1868, there were now four levels: precise, primary, secondary, and tertiary, with the first three characterized by average triangle

closing errors of "about" 1, 3, and 5 seconds, respectively.[39] Under this new schema, *precise* was the highest level, "equal in accuracy [to] *primary* .... as previously designated," and what was now the primary network was "used principally as a means of connection between the precise and secondary work." Although Sutherland dismissively noted that "no tertiary triangulation is contained in this report," the new hierarchy recognized the need to tie additional points and smaller triangles to the secondary triangulation to accommodate public works surveys and other engineering projects.

By the end of the twentieth century, electronic distance measurement (EDM) was supplementing triangulation with trilateration for geodetic and engineering surveys. *Trilateration* uses distances rather than angles to solve the intersection problem in figure 1.4. Post–World War II advances in electronics took advantage of a simple principle: launch a pulse of light at a distant reflector, measure the difference in time between the pulse's emission and return, and use the speed of light to calculate the distance. Advanced EDM instruments with errors less than one part in 200,000 exceeded the requirements for first-order surveys by a factor of 10.[40] By century's end, third-order surveys were typically based on the less expensive total station, which combined a short-range EDM unit with a theodolite. Trilateration is also a fundamental principle of the global positioning system (GPS), widely used for vehicle navigation and scientific data collection as well as an indispensable element of the mobile phone and smartphone.

But wait, there's more. Able to calculate geographic positions to within millimeters—extreme accuracy requires multiple readings—GPS afforded independent verification of nineteenth-century baselines. Although the National Geodetic Survey (NGS), which inherited geodetic responsibilities after the Coast and Geodetic Survey was folded into NOAA in the 1970s, has moved well beyond the legacy networks of Hassler, Cutts, and Schott, the Eastern Oblique Arc caught the fancy of a cadre of geocaching enthusiasts, who hide or recover buried objects identified by their GPS coordinates. In spring 2008, for instance, a GPS hobbyist known as Papa-Bear-NYC sought out forgotten monuments of the New England geodetic network and posted his findings online.[41] Equipped with a handheld GPS receiver, a metal detector, a camera, and a list of published coordinates, he located several monuments of the original Massachusetts primary network, including the North Base station, hidden under the fall leaves in a backyard in

Foxboro, Massachusetts, where the home's owners allowed him to search. Papa Bear's finds included the original station monument (figure 1.12), a "3/4" copper bolt set in a 6" square granite block," and its two reference markers, placed nearby to help surveyors find, or "recover," the station marker. After finding one of the reference markers with his GPS and metal detector, he "guessed" accurately the location of the second, and then located the primary mark over which a tower had been erected long ago.

Figure 1.12. North Base station monument as recovered in 2008. Detail inset (*lower left*) shows a 3/4 in. copper bolt in a 6 sq. in. granite post. Courtesy of Richard Garland.

Papa Bear, whose real name is Richard Garland, reported his find to the NGS, which added "Recovery Note by Geocaching 2008 (RG)" to the station's official data sheet.[42] In addition to reporting that the three markers "were recovered in good condition," Garland included the address (42 Summer Street) where he found the station in the backyard "about 6 feet behind (north) [of] an ornamental wood fence which separates the lawn from the wooded area." He didn't bother looking for the Massachusetts South Base station because its data sheet reported that the Massachusetts Survey had failed in 1934 to find any of the markers

placed ninety years earlier. "Inhabitants in the vicinity .... mentioned that the station had been looked for several times by surveyors in recent times but without success."[43] Apparently, the markers had been destroyed decades earlier, when the New York, New Haven, and Harford Railroad added a third track. The level grade that had made the eleven-mile baseline possible eventually undermined its recovery.

Geodetic survey monuments had a better survival rate in rural Maine than in suburban Massachusetts. Garland was pleasantly surprised that fall when he headed east to the Epping baseline and bagged all five stations on his list.[44] He found largely intact the 12-foot-wide Base Line Road constructed across the Epping Blueberry Barrens by Coast Survey superintendent Alexander Dallas Bache (1806–67) to provide a straight path for the 5.4-mile baseline, measured in seven days in July 1857 (see figure 1.8).[45] At opposite ends of the road, he recovered station monuments with copper bolts in granite, both "in good condition."[46] In his geocaching post, Garland paid homage to Bache's original work, noted that the corrected measurement of 8,715.9422 ± 0.0158 m "was calculated to be on the order of 1 part in 500,000," and gloated in boldface that "recent GPS measurements actually confirm this impressive accuracy."[47]

When I asked for clarification, Garland pointed me to Harry Nelson, the senior geodesist at the Maine Department of Transportation, who provided a *Bangor Daily News* account of how he helped recruit 43 other surveyors, who assembled in eastern Maine on the third Saturday in October 1991 with nine civilian GPS units—the only ones in the state—to estimate coordinates for thirteen points, four of them along the Epping baseline.[48] Multiple units were needed because a US Department of Defense policy known as "selective availability" blurred readings by nonmilitary receivers.[49] But multiple receivers, at least one at a station with precisely known coordinates, could filter out the noise.[50] The resulting unblurred coordinates, apparently never published, were used to calculate a length "within a centimeter" of Bache's original measurement.[51] Impressive.

When I asked my contacts at the NGS for comment, they offered nothing more definitive than the official data sheets for the endpoint stations, Epping East Base and Epping West Base, which report that "horizontal coordinates were established by GPS observations and adjusted by the National Geodetic Survey in June 2012."[52] Although GPS cannot measure distances directly, it provides coordinates that can be used to compute distances. The NGS also provides online software, with which I calculated the distance between endpoints as 8,715.9055 m—and although Bache's measurement of 8,715.9422 m is more than twice the estimated probable error (± 0.0158 m) above my GPS-based calculation, it's strikingly close nonetheless.[53]

Geodesy has numerous stories, most framed by networks such as those I explore in chapter 2, "Geometry." There's the electronic telegraph network, which helped the Eastern Oblique Arc propagate reliable estimates of longitude throughout the region. There's also the Great Arc of the 39th Parallel, which spanned the country from the Atlantic to the Pacific and joined triangulation arcs reaching from Mexico to Canada to inform a mathematical description of the planet as a sphere flattened at the poles and wider at the equator. Though the earth is not a perfect sphere, a simplified geometric representation, known as a *spheroid* or *ellipsoid*, provides a unified frame of reference for latitude, longitude, and elevation. And the intercontinental triangulation network that vaulted the Atlantic to link North America to Europe and Asia was an endeavor facilitated by dynamic networks of geodetic satellites, including giant reflective balloons that served as moving trilateration targets.[54] And finally, there's the role of the military, which recognized the importance of geodesy and global referencing systems in an era of intercontinental ballistic missiles (IBMs) and cruise missiles. This pursuit of precision also informed studies of crustal dynamics, continental drift, and sea level rise.

In chapter 3, "Symbols," the focus shifts from measurements to graphics, in particular the representation on paper and electronic screens of transportation systems and other surface networks. After mapping evolved to encompass canals, roads, and railways, topographic and thematic maps became more diverse and pervasive, as exemplified by pictorial and abstract symbols for subways and airlines. Chapter 4, "Infrastructure," examines the role of maps in planning, building, and managing networks that move goods, people, and energy and highlights the use of maps in keeping track of and protecting buried utility lines.

Chapter 5, "Telecommunications," focuses on nineteenth-century telegraphy, which not only facilitated efficient longitude measurements at triangulation stations but also fostered the efficient accumulation of atmospheric measurements used in making the synchronous weather map. Getting the data together at a

single location where nationwide measurements could be plotted, and plotting them several times a day, was essential to useful forecasts of severe storms. As the twentieth century progressed, national and global telecommunication networks not only distributed timely news maps but also gave the world's weather services a framework for making and distributing Doppler radar maps, now a staple on the internet along with cartographic predictions that look days, and even months, into the future. These stories remind us that disseminating geographic information is as much a part of cartography as collecting and mapping data.

Chapter 6, "Topology," turns to computer-readable networks known as *geographic base files* and *data models*, which use binary digits to describe relationships among households, businesses, and geographic features such as city blocks and street addresses. Among the earliest was the address coding guide (ACG), with which the US Census Bureau tabulated data scanned from mailed questionnaires for the 1970 Census. Later developments fostered a digital, data-driven geography that included interactive maps of census data, minimum-distance maps for dispatching ambulances and fire engines, GPS-based navigation systems, and efficient configurations of stores and distribution warehouses for large retail businesses eager to minimize transportation cost. Chapter 7, "Control," summarizes the diverse relationship between maps and networks as well as the role of electronic maps in guiding driverless cars and promoting partisan gerrymandering. Like many versatile tools, maps can serve nefarious aims.

2

GEOMETRY

Once a baseline has been carefully measured to determine its length, we must then locate it on a world map. Although the baselines and networks of surveyed triangles produced starting in the nineteenth century eventually helped cover the country with topographic maps at various scales, these maps could not be anchored to specific latitudes and longitudes without the locational frameworks examined in this chapter. It introduces geometric concepts that link geodetic surveying and mapmaking to astronomy, geophysics, GPS satellites, continental drift, and modern warfare.[1] As a chapter title, "Geometry" not only reflects the word's etymology (earth + measurement) but also relates the abstract three-dimensional model that geodesists call the *figure of the earth* to a set of precise horizontal or vertical measurements known as a *datum*. If you're fascinated with the details that make modern cartography a distinctive and indispensable technology, you'll love this chapter—at least, I hope so. But if you're put off by details, you might hate it, though I hope not. After all, modern cartography encompasses not only intriguing pictorial maps, intellectually stimulating political atlases, and zoomable maps that help you get from *A* to *B* but also covers a comparatively complex mathematical foundation analogous perhaps to the agricultural underpinning of dining out.

Bridging cartography's humanistic and STEM—that is, science, technology, engineering, and mathematics—sides is a network of sorts called a *graticule*, defined as a lattice of parallels and meridians with a constant increment. On a 10° graticule, for example, parallels in the Northern Hemisphere increase in steps of 10°, from 0° at the equator, to 10°N, 20°N, and so on, to 80°N, just shy of

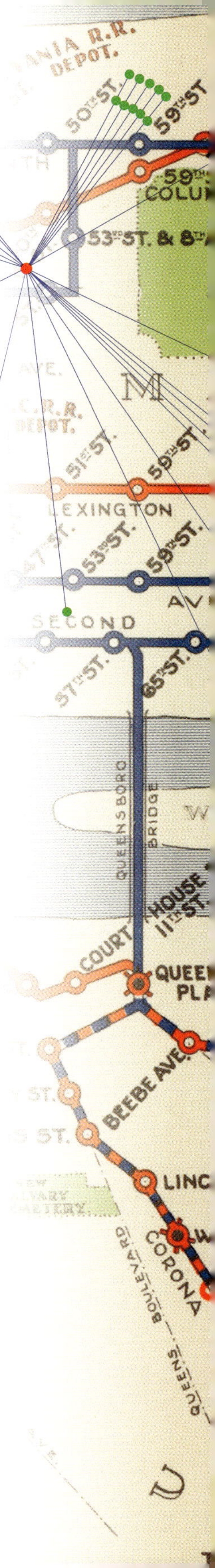

the North Pole at 90°N (a point on some map projections and a line on others). Meridians, which run from pole to pole, are similarly separated. A graticule is common on small-scale world maps, where the increment is large, as well as on large-scale maps, where the increment might be only a minute (one-sixtieth of a degree) or less. The first graticule to appear on a map might have been put there by the Greek geographer and mathematician Eratosthenes (ca. 275–194 BC), who used parallels and meridians to divide a map into quadrangles.[2]

The most basic of these measurements is *latitude*, defined simply as a point's angular distance north or south of the equator, shown as angle A in figure 2.1. Geometry tells us that latitude can be measured indirectly as the angle of elevation A' of the polar star above the horizon. This substitution assumes that an observer has no trouble viewing the bright star Polaris through a cloudy night sky and correctly discerning the horizontal plane, which is perpendicular to a line running directly outward from the center of the earth. I drew this cross section of the earth with the observer's position at the top, because in the mid-latitudes, where most of us live, the polar star will lie off to the side, well away from a *plumb line* (a string attached to a weight with a pointed tip) directed downward toward the earth's center. Because Polaris is over four hundred light-years away, the line of sight from P is essentially parallel to the earth's axis and thus perpendicular to the plane of the equator. If you recall your plane geometry, you can prove that angle A' (which you can measure) equals the latitude angle A (which you can't, at least not directly).

There's a hitch, of course. Measuring the precise latitude is complicated by having to wait until dark on a mostly cloud-free night and also correct for the possibility that Polaris might be nearly a degree off the polar star's ideal position directly above the North Pole.[3] Relative to the earth's axis, Polaris is continually moving, but because this movement is predictable, tables published in an ephemeris provide an appropriate correction for any day of any year. Although GPS made this conceptually important measurement unnecessary, in 2015 the US Navy resumed training its officers in celestial navigation, once a lost art.[4] A battalion of anti-satellite weapons could cripple the satellite navigation system, but more plausible scenarios include hackers as well as a massive solar storm such as the Carrington event, which disrupted electric telegraphy in 1859.[5]

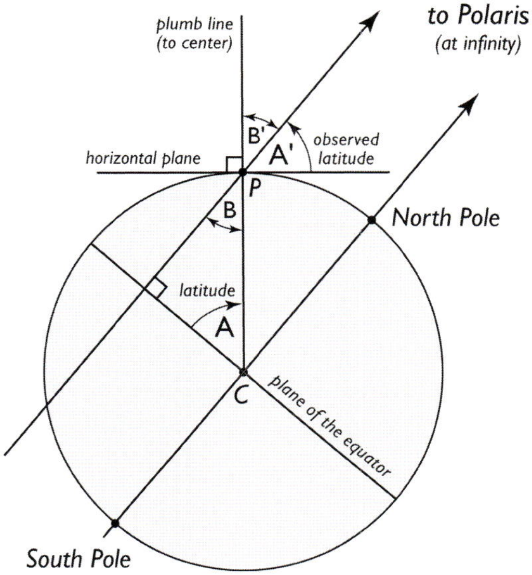

Figure 2.1. Geometric relationships between latitude (angle *A*) at point *P* and the elevation (angle *A'*) of the polar star above the horizon.

Two further complications are gravity and a nonspherical earth. Because the density of mass below the earth's surface varies noticeably from place to place, a plumb line will not necessarily point directly to the center of the earth. Because of this deflection of the vertical, surveyors cannot assume that the horizontal plane is perpendicular to a plumb line. Gravity variation also affects the spirit levels used to set up a plane table or theodolite. Moreover, because the earth is flattened at the poles and wider at the equator—in other words, not a perfect sphere—straight lines perpendicular to the horizontal plane at different locations will not meet at any single point.

Cartographers and geodesists keep calculations manageable with a simple mathematical model called an *ellipsoid*, which is a three-dimensional figure like a sphere but based on a cross section that is an ellipse rather than a circle. Because this ellipse's minor (shorter) axis coincides with the earth's axis while its major (longer) axis lies in the plane of the equator (figure 2.2, *upper left*), rotating this elliptical cross section about the shorter axis yields a three-dimensional ellipsoid. In contrast to the sphere and its circular cross section, which share a single radius,

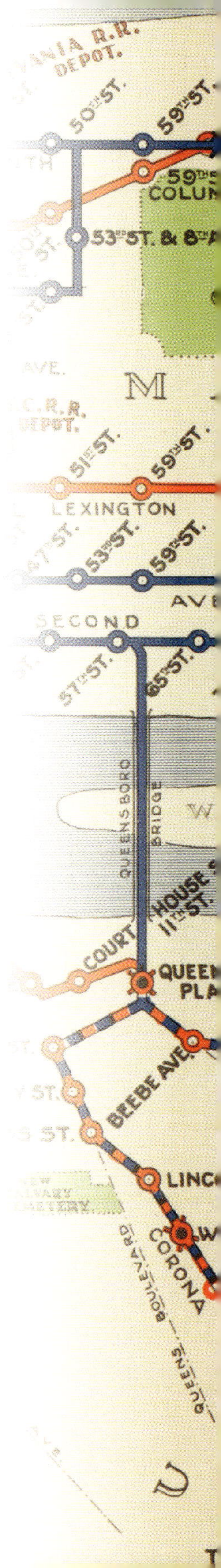

the ellipsoid and its cross section share two radii: a semimajor axis *a* about 1/300 longer than the semiminor axis b. The obvious exaggeration in figure 2.2 is necessary to describe how this small degree of flattening affects relationships between the ellipsoid, the plane of the equator, and the center of the earth.

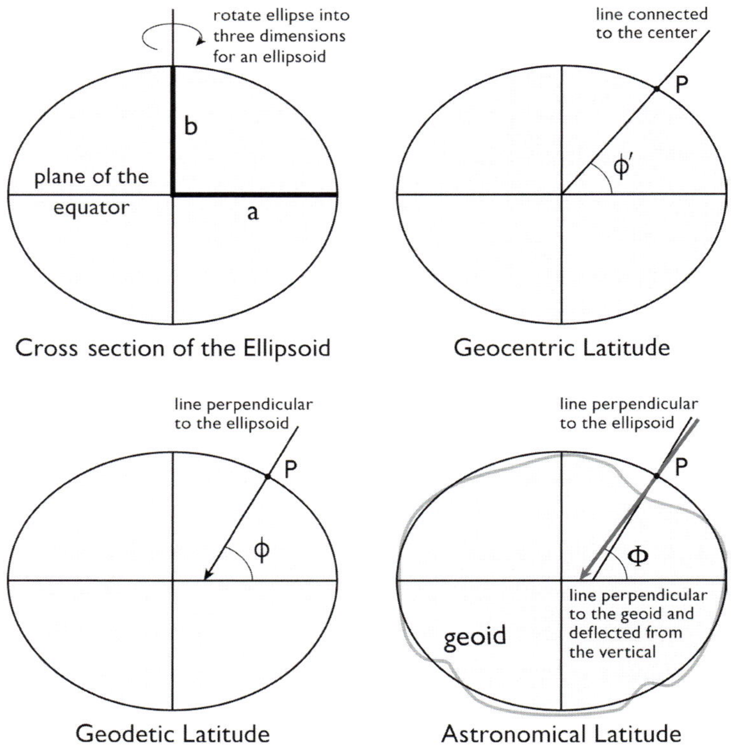

Figure 2.2. Cross sections of the ellipsoid and various definitions of latitude. The flattening of the ellipsoid and the separation between geoid and ellipsoid are exaggerated for clarity.

Figure 2.2 also illustrates three different definitions of latitude. At the upper right, the angle between the equatorial plane and a straight line from point P to the center of the ellipsoid describes *geocentric latitude* (ϕ'), akin to simple spherical latitude but defined for the ellipsoid. At the lower left, the angle between the equatorial plane and a line perpendicular to the ellipsoid at point *P* describes the *geodetic latitude* (ϕ), used to position parallels on topographic maps and listed in tables of spherical coordinates.

A third type, described at the lower right in figure 2.2, is *astronomical latitude* (Φ), defined as the angle between the equatorial plane and the plumb line, also called the *true vertical* or the "line of action of gravity at the station."[6] The cross section includes a hypothetical surface known as the *geoid*, which provides a worldwide zero reference for elevation. Think of the geoid as the surface that would result if mean sea level could be extended without tides through the continents. The geoid undulates around the globe in response to variations in density of the land mass and is thus relevant to astronomical latitude because gravity deflects the plumb line through *P* away from a line perpendicular to the ellipsoid. To illustrate this effect, my diagram greatly exaggerates the undulations and deflection.

Nineteenth-century geodetic surveyors treated the discrepancy between astronomical and geodetic latitude as a "station error" and calculated adjustments on the basis of a triangulation network, the parameters *a* and *b* of the ellipsoid, and statistical averaging.[7] Although this discrepancy was seldom more than a few seconds of arc—a single second is only one-sixtieth of one-sixtieth of a degree—station error was hardly negligible to scientists who measured baselines to within several centimeters.

In addition to being mathematically savvy, geodesists had to be well versed in astronomy. Because key stars have predictable motions published in tables that relate their positions to latitude and day of the year, geodesists calculated latitude and longitude using angular measurements of a star's position relative to a vertical line, the horizon, or another celestial body. In the index to his 1898 *Text-Book of Geodetic Astronomy*, John Fillmore Hayford (1868–1925) listed five different methods for measuring latitude and five different methods for measuring longitude.

A geodesist and computational expert at the Coast and Geodetic Survey, Hayford also described several specialized instruments for measuring latitude, notably the *zenith telescope*, designed to point upward and measure the precise angle between the vertical (zenith) at the location and the line of sight to a particular star.[8] When this angle attained its minimum for the day in question—which coincided with the star's highest altitude above the horizon—the star was said to have "crossed the meridian," and the angle was recorded as its zenith distance. The star was paired with another star expected to cross the meridian a short time

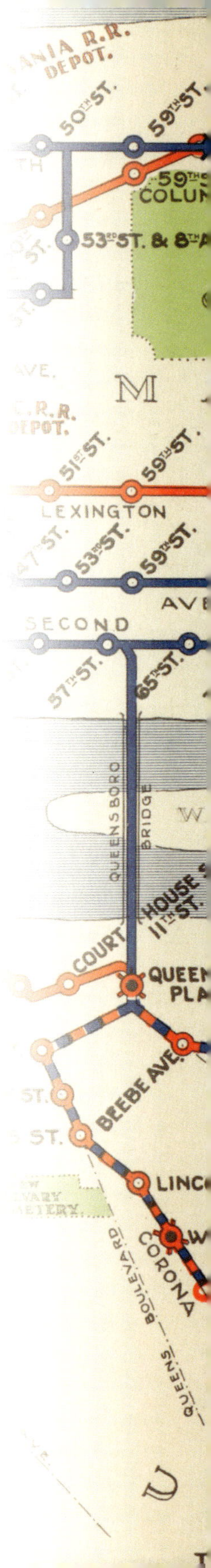

thereafter on the opposite side of the observer's zenith. Because one star was to the north of the zenith position and the other star to the south, the zenith distances of the two stars could be added to obtain a zenith difference useful in calculating latitude. And because multiple zenith differences were needed to dampen observation errors, the night's activities were guided by a carefully prepared "observing list" of paired stars, to be measured in sequence. The list included each star's catalog number, magnitude, anticipated location, and the approximate time of its meridian crossing.

Astronomical observations used in measuring longitude were events such as a *lunar culmination*, when the moon reached its highest altitude in the night sky, or various occultations and transits involving Earth's moon or the satellites of Jupiter.[9] An *occultation* occurs when one body is hidden behind another, and *transit* refers to a smaller body crossing in front of a larger body and possibly casting a shadow. These events were listed in an ephemeris along with their time of occurrence (sometimes to the nearest decimal second) at an observatory with a known latitude. The difference in local time between the event's calculated (predicted) occurrence at Washington, DC, or Greenwich, England, and when the event was observed at a geodetic station could be converted to a difference in longitude because the earth rotates through 360° in twenty-four hours and 15° in one hour. For example, if an event predicted to occur at observatory *A* at exactly four hours past midnight local time was observed at station *B* at two hours past midnight local time, the two-hour difference in local time represents a 30° difference in longitude. Therefore, if the longitude at *A* is, 77°0'12" W, the longitude at B would be 107°0'12" W. Sketch an example or two on paper, to make certain you understand the principle.

Estimating a location's latitude and longitude was a demanding process of measuring angles, observing times, and calculating differences, and an actual calculation was never straightforward. A geodesist first determined the location's apparent solar time, based on the sun reaching its highest point in the sky that day at noon. He would then use the so-called "equation of time" to convert to local mean time, also called *mean solar time*, which can differ by as much as sixteen minutes from apparent solar time.[10] This difference, which varies throughout the year, reflects both the 23.4° axial tilt of the planet and the elliptical (rather than circular) shape of its orbit around the sun. An adjustment was necessary because ephemerides published only the mean time for specific events.

Several sources of uncertainty undermined the use of a sextant and a precision timepiece called a *chronometer* to determine apparent local time. A *sextant* is a handheld device for measuring the angle between two lines of sight, say between the sun and the horizon. Sources of error include atmospheric refraction (bending) of lines of sight and the quality of the instrument and its telescope. Although tables showing the measured effects of atmospheric temperature and pressure on the bending of light allowed a partial correction for refraction, which was not always steady, the value of these tables depended on reliable readings of temperature and pressure.[11] No less significant were "observer's errors" related to experience and personality.[12]

Personal bias was particularly troublesome because the variable brightness and definition of stars and satellites affected the visual perception and recorded timing of culminations, occultations, and transits. As Hayford noted, "Every experienced observer, though doing his best to record the time of transit accurately, in reality forms a fixed habit of observing too late, or too early, by a constant interval."[13] Because readings were attributed to individual observers, the Coast and Geodetic Survey (C&GS) not only calibrated an "absolute personal equation" for each observer but also a "relative personal equation" for pairs of observers who worked together frequently but at different locations. According to Hayford, experienced observers typically differed by a quarter second or more, and occasionally by as much as 1.2 seconds.[14] Although a quarter second might seem trivial, the earth's rotation through 15° in an hour equates to 15 arc seconds of rotation in a second of time, or 3.75 arc seconds in a quarter second. On the ground at 40°N, that's about 300 ft.—hardly trivial to geodesists eager to reduce baseline error to a couple of centimeters. As Hayford concluded, "The difficulty of accurately observing these phenomena (including the eclipses) makes the derived longitudes only rough approximations."[15]

Although an ephemeris, a calendar, the positions of heavenly bodies, and a clock adjusted to the most recent sunset afforded a rough estimate of longitude at sea, mariners typically required a precision timepiece calibrated with local time at Greenwich. Reliable chronometers were not available until the late eighteenth century, and then thanks only to the perseverance of John Harrison (1693–1776),

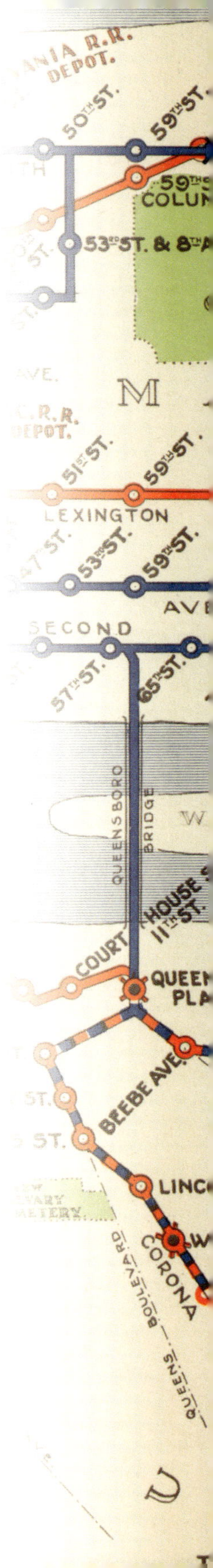

the subject of Dava Sobel's epic biography *Longitude: The True Story of a Lone Genius Who Solved the Greatest Scientific Problem of His Time.*[16] Although precision clocks were also useful on land, Hayford's term, "transportation of chronometers," underscored the problems posed by bumps and temperature fluctuations during overland travel and the consequent need to average readings from multiple instruments. For example, an 1894 effort to determine the longitude difference between Anchorage Point and Sitka, Alaska, employed nine chronometers.[17]

For more efficient and reliable determinations of longitude, geodetic surveyors came to rely on the network of commercial telegraph stations that evolved quickly over the half century after Samuel F. B. Morse (1791–1872) successfully connected Baltimore and Washington in 1844. Experiments in the late 1840s and early 1850s led to what European geodesists called the *American method*.[18] A key instrument was an *astronomical transit* (figure 2.3), cemented to a large stone block for stability so that its powerful telescope, with a focal length of roughly 40 in. and 100 × magnification, remained within the station's vertical north-to-south plane.[19] The observer would elevate the telescope to await the precise moment a particular star crossed the local meridian, represented inside the instrument by a thin wire or etched line across one of the lenses.[20] Meridian crossings were more accurately timed than culminations or occultations.

The apparatus included an electrical circuit that connected a chronometer (precise clock), a chronograph, a battery, and a break-circuit key. The *chronograph* (figure 2.4) consisted of a clock-driven cylinder that rotated once a minute and a pen mechanism that moved parallel to the cylinder's axis, at a uniform rate of a tenth of an inch each minute, to trace a helix on a paper sheet mounted on the cylinder.[21] Every one or two seconds, the chronometer interrupted the circuit and the pen added a small but noticeable jog to the helix, except at the beginning of each minute. This uniform pattern of offsets provided a scale for estimating, perhaps to within a tenth of a second, an event recorded when the observer used the key to break the circuit at the moment a star began or completed its transit across the meridian. The moment of crossing was readily estimated by averaging the beginning and ending times.

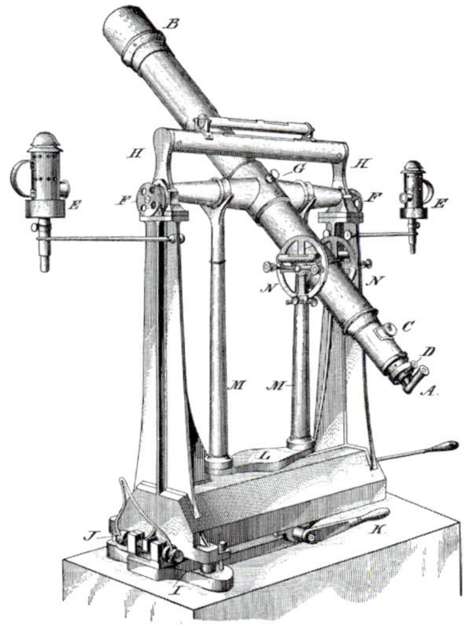

Figure 2.3. Astronomical transit, for observing the exact time a star crossed the meridian. Lanterns *E* at opposite ends of the telescope's horizontal axis provide illumination at night. From John Fillmore Hayford's 1898 textbook on geodetic astronomy.

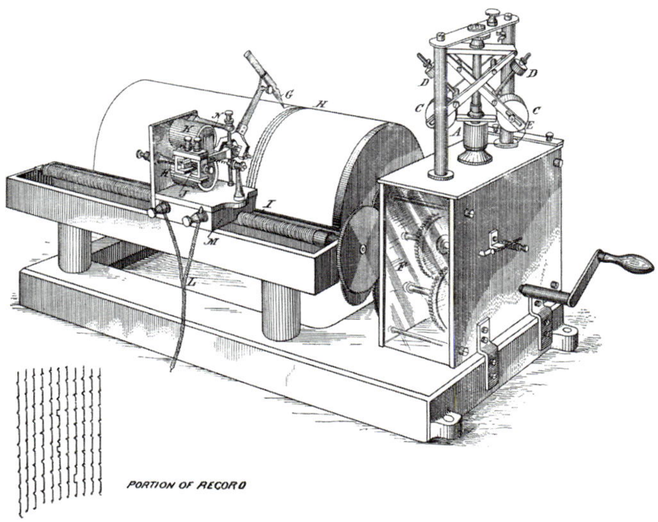

Figure 2.4. Chronograph, for relating the time of the meridian crossing to the exact local time. The handle winds the spring-driven clock that rotates the drum and drives the pen mechanism parallel to the drum's axis. *Lower left*: a portion of the helix inscribed by the pen, connected electrically to a chronometer and an observer's key. From John Fillmore Hayford's 1898 textbook on geodetic astronomy.

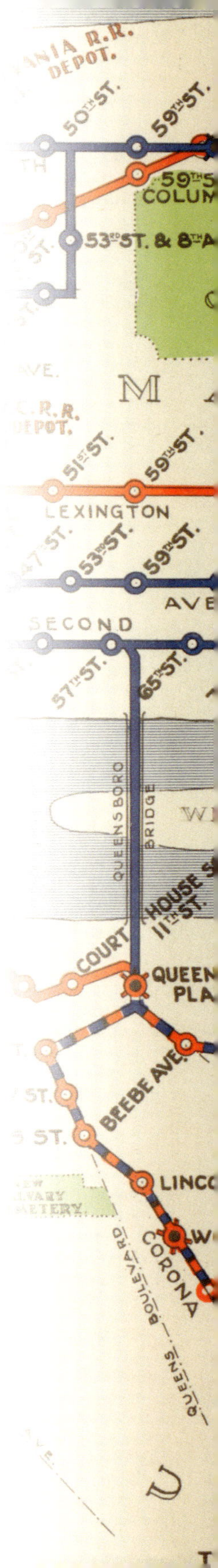

To confirm that the chronometer and chronograph were operating properly, the observer ran the apparatus offline, as described by the circuit diagram in the upper half of figure 2.5, which includes a telegrapher's key and sounder for communicating with a second observer at a distant station with a similar apparatus. Using Morse code, the observers discussed sky conditions that might interfere with their work and planned their evening session, generally consisting of two sets of ten stars each, starting after 10:30 p.m., when commercial traffic was minimal.[22] Between the two sets, they exchanged additional signals to compare their respective chronometers. When they were ready to exchange signals and the telegraph company gave them a clear line, the observers' local circuits were connected to the main telegraph line, as described in the lower half of figure 2.5. The differences in time between when the two observers recorded the same event became estimates of their difference in longitude, and averaging these differences after adjusting for observation errors yielded a relatively reliable measurement.

Multiple stars observed on five nights were used to identify and adjust observation errors.[23] To compensate for their personal equations, the observers would then exchange places, leaving each station's apparatus in place, and observe the same sets of stars for another five nights.[24] This often meant a day or more of travel for both observers. According to Hayford, "eliminating the personal equation" typically reduced the error for ten sets of observations to less than 0.2 seconds of time, or 3 arc seconds of longitude. "To whatever these errors may be due," he reasoned, "they seem to be fairly well eliminated from the mean for the station." When longitudes were integrated with the triangulation network, he noted, "the discrepancies in closing the various 'longitude triangles' are always less than 0.10 seconds."[25] In listing geographic coordinates, the C&GS typically reported seconds of latitude and longitude to one, or sometimes two, decimal places.

Telegraphic determination of longitude was well established in 1878, the publication date of the annual report of the C&GS for 1876, which included a map of telegraphic connections for determining longitude.[26] The map covered more than three decades, beginning with a pioneering exchange of signals between Washington and Philadelphia in October 1846 and extending through the end of 1877, when the network of paired stations included San Francisco, San Diego, and Seattle on the Pacific coast as well as New Orleans, Galveston, Texas, and Key West, Florida, on the Gulf of Mexico.[27] According to a note at the bottom of

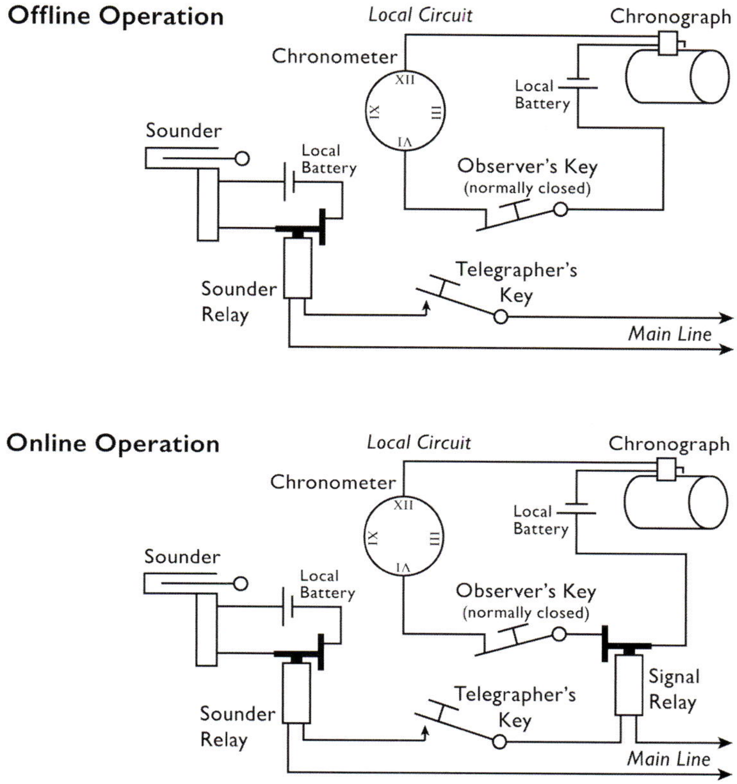

Figure 2.5. Circuit diagrams describing offline (*top*) and online (*bottom*) operation of a chronograph connected to a telegraph line. Redrawn from Hayford's 1898 textbook on geodetic astronomy.

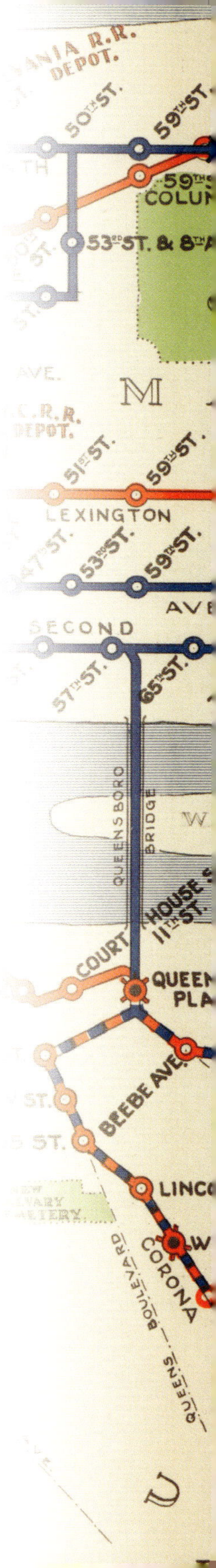

the map, during this period ninety-four longitudes were "determined by means of electric telegraph."

Figure 2.6 is my much-reduced generalization of a map printed with black and blue inks on a sheet slightly more than two feet wide. My cartographic summary omits state names and boundaries as well as the names of most longitude stations but accurately captures the network's extent and connectivity as well as key design details of the original.

Because longitude was determined for pairs of stations, parts of the map resemble the hub-and-spoke pattern of airline route maps that emerged in the 1970s, after Congress deregulated the industry and commercial carriers abandoned multistop, up-and-down trips that mimicked passenger trains. Like an

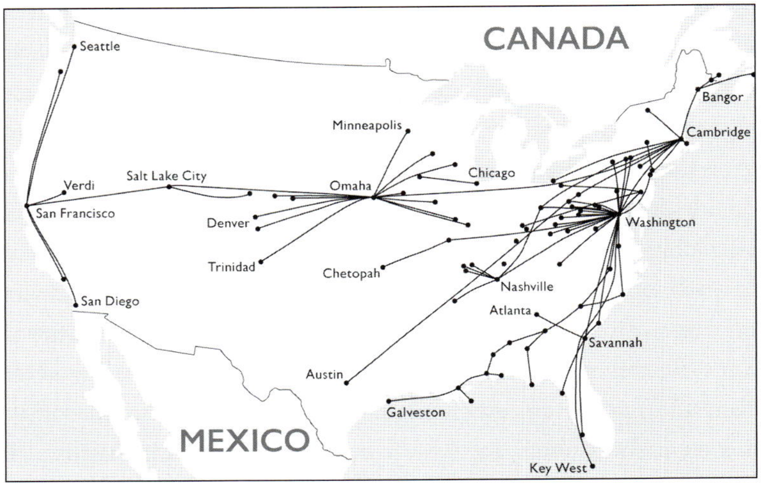

Figure 2.6. Longitude stations and connections determined by the electric telegraph between 1846 and 1877, adapted from a much larger 1876 C&GS map. This generalized version omits the connections to Europe completed in 1866, 1870, and 1872 and shown in figure 2.7. Three unconnected dots in Kentucky (northeast of Nashville) were linked to Cambridge, Massachusetts.

airline map, the telegraphic-connection map is concerned largely with pairs of endpoints, rather than with the routes of telegraph lines, which typically followed railways. Curved (rather than straight) lines reduce graphic conflict, as exemplified by the line connecting Cleveland, Ohio, with Cambridge, Massachusetts, which bends northward to avoid the symbol for a station near Binghamton, New York. Nonetheless, clutter is inevitable because of the large number of comparatively short links converging on Washington, DC, home to C&GS headquarters and the Naval Observatory. Because this convergence separates three stations in central and northern Kentucky from the Harvard Observatory, their common partner in Cambridge, Massachusetts, the original map replaced their connecting lines with the phrase "to Camb." aligned with a short arrow pointing northeast. Look carefully, and you will notice that I left the three dots unconnected: an awkward solution excused as cartographic license.

As appropriate for determinations of longitude, the connections run largely east–west. Notable exceptions include north–south connections through San Francisco to serve a triangulation network initiated along the Pacific coast in 1850, a similar extension down the East Coast to the tip of Florida, and links to various points on the Eastern Oblique Arc, which stretched from eastern Maine to New Orleans (see figure 1.9 in chapter 1). According to Charles Schott, who hailed the arc's completion in his 1902 report, "The latitudes depend almost exclusively on observations with zenith telescopes, and the longitudes on telegraphic transmission of time."[28] That terminal points such as Chetopah, Kansas; Trinidad, Colorado; and Verdi, Nevada, were insignificant cities reflects the goal of anchoring the nation's first-order triangulation network rather than serving centers of population.

The long, sparse connection from Omaha through Salt Lake City reflects the path of the first transcontinental telegraph, completed in 1861, eight years before the "Golden Spike," hammered into place at Promontory Summit, Utah Territory, on May 10, 1869, to signify completion of the transcontinental railroad.[29] The original Coast Survey had initiated a triangulation network in California in 1850, and Superintendent Alexander Dallas Bache eagerly awaited a telegraphic connection between San Francisco and Washington. In his annual report for 1861, he praised the telegraph "as the most admirable and simple means of determining the difference in longitude between these points" and confidently predicted "the intelligent president of the line will give us every facility for this purpose."[30] But the Civil War (1861–65) intervened, and San Francisco was denied a telegraphic longitude until 1869, when Coast Survey assistant George Davidson (1825–1911) obtained permission from city officials for an observatory in Washington Square and arranged a direct connection to the Western Union network.[31] Between February 15 and April 4, Davidson exchanged clock signals with the Harvard Observatory in Cambridge on twelve nights, with Salt Lake City on twelve nights, and with Omaha, Nebraska, on eight additional nights. The longitude network's metaphorical golden spike beat out its railroading counterpart by almost three months.

Davidson had helped secure another final link: the telegraphic longitude connection through Calais, Maine—less than 50 miles from the Epping baseline—with the pioneering transatlantic cable completed in 1866. Figure 2.7, a facsimile of the upper-right corner of the 1846–77 map summarized at a much smaller size in figure 2.6, describes the Calais link along with two other connections to Europe, completed in 1870 and 1872 through Duxbury, Massachusetts. Davidson had been a Coast Survey field worker in California from 1850 to late

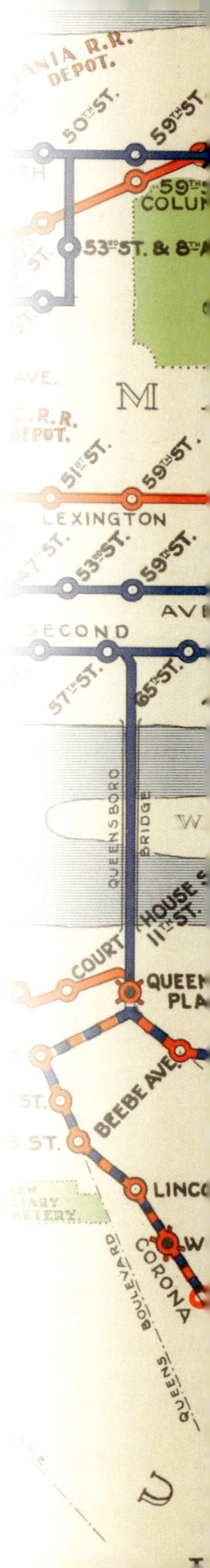

1860, when the South's secession from the Union seemed imminent. Bache, who was a member of the clandestine Blockade Strategy Board, brought him back east, to plan defenses against the Confederate navy and make hydrographic charts relevant to a blockade of the southeastern coast.[32] Bache had taught Davidson astronomy at Philadelphia's Central High School in the early 1840s and was well aware of his former student's skill as an astronomer, engineer, and geodetic surveyor.[33] The war's end found Davidson in England, on an emergency assignment to help Bache, whose health had deteriorated while negotiating a telegraphic exchange with the Royal Observatory.[34] In September 1866, Davidson was sent to inspect telegraph lines between Calais and the western terminus of the transatlantic cable, in Newfoundland, and to set up astronomical instruments at Calais, where a longitude station established in 1857, in anticipation of an undersea link to Europe, had fallen into disrepair.[35]

Bache had been intrigued by the possibility of a telegraphic longitude connection with Greenwich since the early 1850s, when Cyrus Field's Atlantic Telegraph Company began raising money for a cable across the Atlantic seafloor.[36] Completed in 1858, the cable failed after only a month's operation.[37] A second attempt, in July 1865, was thwarted when the cable snapped at midocean just off the stern of the massive cable-laying ship *Great Eastern*. In July of the following year, the ship's crew laid a new cable, from Valentia Island, on the west coast of Ireland, to Heart's Content, on the Newfoundland coast. They then returned to the area where the 1865 cable had snapped, recovered the lost end, spliced it onto fresh cable, and steamed back to Heart's Content, which now had twin telegraph lines to Europe.

A working undersea cable was just one step in connecting the Royal Observatory with longitude stations in the United States. Davidson found telegraphic communications between Heart's Content and Calais unreliable because of poor insulation on the roundabout 1,100-mile line between the two stations, which included several repeater stations, necessary because electrical resistance weakened signals sent over a long wire.[38] Repeaters allowed a long circuit to be split into two or more shorter circuits; at each junction, an automatic device or human operator received a weakened signal and transmitted onward a stronger signal with the same message. Although the link's four automatic repeaters were deemed acceptable sources of added error, the Coast Survey provided

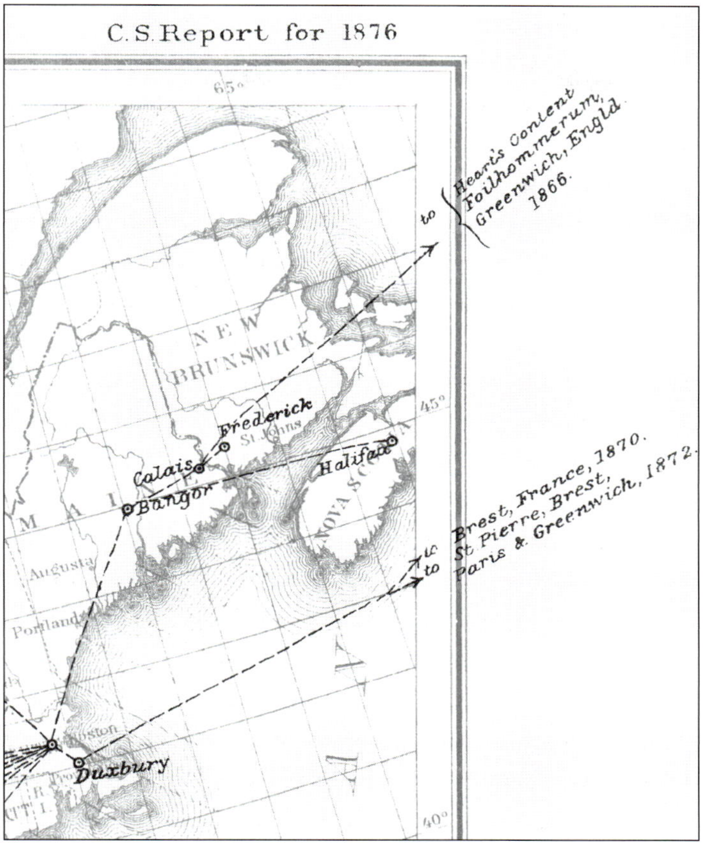

Figure 2.7. Facsimile of *upper-right* corner of the 1876 Coast Survey map of longitude stations and connections determined by electric telegraph.

battery-powered repeaters for the two manual repeater stations, at which received messages previously had to be written out and rekeyed. Even so, experimental transmission was disappointing for seven weeks until December 11, 1866, when "suddenly," as Coast Survey assistant Benjamin Apthorp Gould (1824–96) described in the annual report of 1867, "the long desired communication was found to be established," apparently because "a sharp frost had thrown the otherwise defective line into a condition of admirable insulation, so that an interchange of clock signals [could be] effected without difficulty."[39] On the other side of the Atlantic, the telegraphic exchange between the Coast Survey's astronomical station at Foilhommerum on Valentia Island and the Royal Observatory at

Greenwich depended on almost 600 miles of landline as well as cables across the straits of Valentia (about .75 mi.) and between Ireland and Wales (about 60 mi.).[40]

Gould, who directed the Coast Survey's longitude program from 1852 to 1867, had developed a detailed program of observations for the "Trans-Atlantic Longitude Campaign."[41] Because the stations in Newfoundland and Ireland were too far apart for star signals, longitude was determined by comparing carefully calibrated astronomical clocks, adjusted using nightly observations.[42] The protocol for exchanges between Heart's Content and Calais was comparatively straightforward, with clock signals exchanged each night as long as the Heart's Content station was operating, and Calais going first in each two-way set. By contrast, exchanges between Foilhommerum and Heart's Content "depended upon the convenience of the telegraph company" and were generally confined to the interval of 10 p.m. to 6 a.m., Greenwich time, with special codes transmitted to indicate readiness and Foilhommerum generally going first.[43] Instructions also called for "experiments" to test the velocity of transmission in both directions and along the two lines, separately and in tandem.[44] The Coast Survey ran exchanges over these two portions of the overall link, and British astronomers provided the longitude difference between Foilhommerum and Greenwich.

Gould's report discussed various adjustments, corrections for observers' personal equations, averaging, and the suppression of one exceptionally low measurement between Heart's Content and Calais, attributed to "some serious disturbance to the clock."[45] The three remaining estimates for the Heart's Content–Calais segment were weighted equally, but one of the five for the Foilhommerum–Heart's Content segment was assigned a weight of only one-half because a progressive increase between successive sets "possibly indicates a variability in the clock rate." The difference in time units between highest and lowest of the three latter longitudes was only 0.11 second, while the corresponding disparity among the five longitudes estimated for the longer, undersea Heart's Content–Calais segment was an impressively smaller 0.026 second. Gould combined the estimates for the three complete segments as shown in table 2.1.

He then added incremental differences in longitude from Calais to Bangor, Cambridge, New York City, and Washington, DC, and reported the longitude of the Naval Observatory, as $5^{h}\ 8^{m}\ 12^{s}.39$. In the late nineteenth century, longitude was commonly expressed as a time difference with superscripts. Mindful of

Table 2.1. Longitude difference between Greenwich and Calais

Segment	Hours	Minutes	Seconds
Greenwich-Foilhommerum	0	41	33.29
Foilhommerum-Heart's Content	2	51	56.54
Heart's Content-Calais	0	55	37.72
Greenwich-Calais	4	29	7.55

Note: *The longitude difference (in time units) between Greenwich and Calais was calculated by combining the separate determinations for the three connections.*

congressional interest in another Washington landmark, Gould noted that the Capitol Dome was $10^{s}.17$ to the east and reported its longitude as $5^{h}\ 8^{m}\ 2^{s}.22$ west of Greenwich.[46]

Additional transatlantic cables led to further telegraphic determinations. In 1869, a French firm connected Brest, France, with Duxbury, on Cape Cod, Massachusetts, through an intermediate station at St. Pierre, a French possession 16 miles off the coast of Newfoundland.[47] This cable allowed a longitude connection between the Harvard Observatory, at Cambridge, with Brest in 1870, and between Harvard and a European "longitude triangle" linking Brest, Paris, and Greenwich in 1872.[48] Additional transatlantic cables followed, including a link between Canso, Nova Scotia, and Waterville, Kerry (on the west coast of Ireland), in 1884, which provided a longitude determination in 1892 between Greenwich and Montreal, Canada, subsequently integrated with the network of longitude determinations in the United States.[49]

Improved longitude estimates show a small but marked increase. In 1851, for instance, various astronomical observations, including 373 shipments of chronometers across the Atlantic Ocean, had established the longitude of the Harvard Observatory, in Cambridge, as $4^{h}\ 44^{m}\ 29^{s}.50$.[50] The initial transatlantic cable determination in 1866 increased this measurement to $4^{h}\ 44^{m}\ 30^{s}.85$, subsequently refined in 1874 to $4^{h}\ 44^{m}\ 30^{s}.98$.[51] An 1884 report placed the dome of Harvard Observatory at $4^{h}\ 44^{m}\ 30^{s}.993$, and in 1897 its longitude increased again, to $4^{h}\ 44^{m}\ 31^{s}.046$.[52] San Francisco experienced a similar westward shift. An early estimate based on lunar culminations placed Telegraph Hill at $8^{h}\ 09^{m}\ 33^{s}.29$, but by 1855, refinements based on 206 moon culminations as well as chronometer transportations had increased its longitude to $8^{h}\ 09^{m}\ 34^{s}.37$. A greater jump followed in 1869,

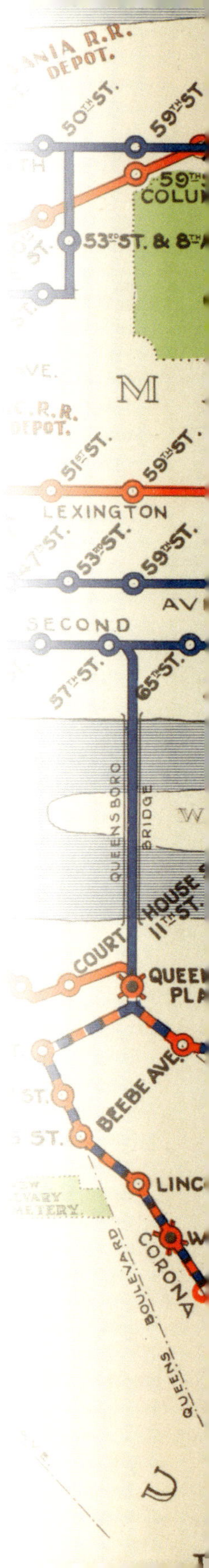

when the transcontinental telegraphic connection with Cambridge pushed the time difference to $8^{h}\ 09^{m}\ 37^{s}.46$. Although subsequent comparisons are thwarted by the actual westward relocation of San Francisco's official longitude station, first to Washington Square Park and then to Lafayette Park, the city's longitude increased along with those of Cambridge and the rest of the nation. As C&GS mathematician Charles Schott observed, "The country was considerably wider than had been known before the advent of the telegraphic method."[53]

Schott's comment was buried in a footnote to his 1897 C&GS report that introduced the term "telegraphic longitude net" to describe "a limited number of selected stations" broadly distributed across the country to provide longitudes for the national triangulation as well as for state capitals and key observatories, and "to give consistency and accuracy to all longitudes of surveys connected with this standard system."[54] An accompanying map reflects (but is not limited to) the country's three transcontinental telegraph routes, one near the southern border connecting San Diego through El Paso, Texas; another below the Canadian border connecting Chicago with Portland, Oregon, and Seattle by way of Minneapolis and Bismarck, North Dakota; and a third in the middle, through Omaha, Nebraska, and Salt Lake City, Utah, to San Francisco. Although my much-reduced, generalized version (figure 2.8) omits a few place names to avoid graphic conflict, the nodes are regionally important cities generally a few hundred miles apart. Also apparent are north–south paths along the Atlantic and Pacific coasts and through key interior junctions, including Nashville, Tennessee, Omaha, and Salt Lake City. On the periphery are Montreal, Calais, and Duxbury, significant for their longitude connections with Greenwich; El Paso and Galveston, as gateways to Mexico and Latin America; and Key West, with undersea connections to Cuba, the Caribbean, and Central and South America.

Not a mere graphic generalization of places with longitude determinations, Schott's map described a working computational network designed to balance the need for broad national coverage with the limitations of hand calculation. Schott's job title, in addition to "assistant," was "chief of the computing division," in which eleven human "computers" solved equations and summarized data for diverse C&GS operations, including geodesy and tidal prediction.[55] As a computational strategy, the longitude net helped reconcile telegraphic differences in time with the triangulation network's baselines and angles and the graticule's

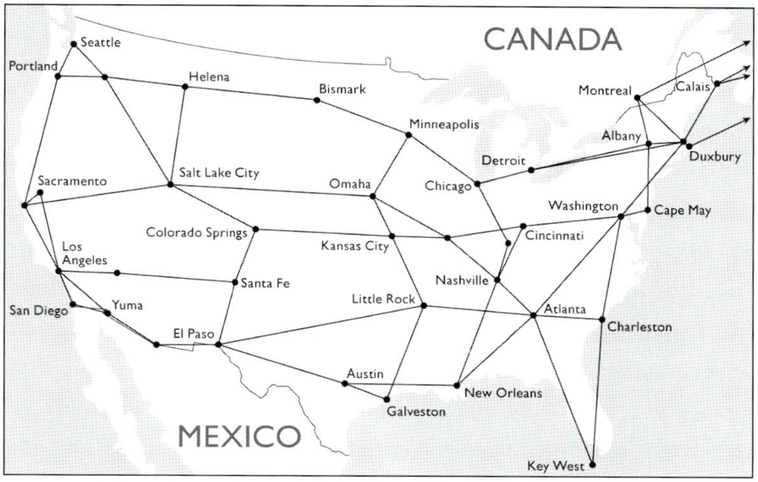

Figure 2.8. Charles Schott's telegraphic longitude net of the United States, which framed adjustment calculations at the end of the nineteenth century.

uniform framework of meridians and parallels, or as Schott so elegantly put it, "to properly disperse in the results the small individual discrepancies, which make their appearance in all fallible measures when confronted with the inherent mathematical conditions."[56]

Another responsibility of the computing division was the map projection tables used to construct frequently used projections, such as the Mercator and the polyconic. An indispensable aid to mapmakers otherwise overwhelmed by trigonometric functions, logarithms, and complex summations, these detailed tables showed the relative spacing or position of grid lines specified in degrees, minutes, and seconds. For large-scale topographic maps and hydrographic charts, projection tables incorporated the ellipsoid believed to best represent the planet's size and shape.

Ongoing least-squares adjustment of the nation's expanding, and ever denser, triangulation network occasionally indicated the need for a new ellipsoid that better accommodated geodetic measurements. Adopting a new figure of the earth was a momentous decision, not taken lightly because of the computational investment in geodetic adjustments and projection tables. Until 1880, geographical positions in the United States were based on an ellipsoid introduced in 1841 by Friedrich Wilhelm Bessel (1784–1846), a German astronomer and

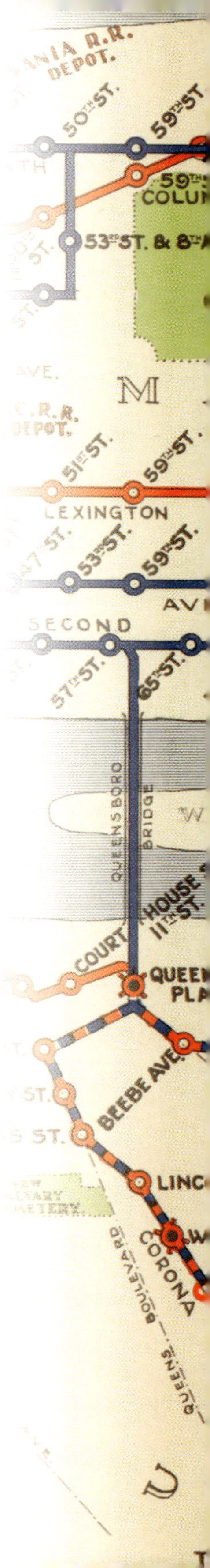

mathematician who calculated the earth's semimajor and semiminor axes as 6,377,397.2 m and 6,356,079.0 m, respectively.[57] An ellipsoid's shape is described by a parameter known as *flattening*, calculated by dividing the difference between the axes by the length of the longer axis and expressing the result as a fraction with one in the numerator, so that the smaller the denominator, the flatter the ellipsoid. Thus, the Bessel 1841 ellipsoid, with a flattening of 1/299.15, is more circular (less flat) than the ellipsoid with a flattening of 1/294.98 presented in 1866 by British geodesist and military engineer Alexander Ross Clarke (1828–1914).[58] With semiaxes of 6,378,206.4 m and 6,356,583.8 m, the Clarke ellipsoid is also slightly larger than Bessel's figure of the earth, based on triangulation data from Europe and central Asia.[59] The Clarke 1866 ellipsoid, which incorporated additional triangulation data from India and southern Africa, replaced the Bessel ellipsoid because it yielded better results for the United States.[60]

These improved results promoted the rollout in 1901 of the United States Standard Datum, a nationwide system of horizontal positions based on a sparse triangulation network framed by the Arc of the 39th Parallel, which connected California with the Eastern Oblique Arc extending from Maine to Louisiana (figure 2.9).[61] The network included shorter triangulation arcs around the Great Lakes as well as offshoots in California southward toward Santa Barbara and in Kansas and Nebraska northward along the 98th meridian. This latter arc, which eventually reached the Canadian and Mexican borders, joined the Arc of the 39th Parallel at Meades Ranch, a geodetic station in Osborne County, Kansas, near the country's geographic center. The US Standard Datum tied this triangulation network to the Clarke ellipsoid at Meades Ranch, thereby providing a common graticule for mapping any part of the country, large or small. In projecting horizontal positions onto the Clarke ellipsoid, the datum established not only latitude and longitude but also the distance relationships essential for making maps at specific scales. In 1913, Canada and Mexico adopted the United States geodetic framework, which then became the North American Datum.[62]

Geodetic surveying is an ongoing pursuit of ever more complete and precise representations of the planet. As the twentieth century progressed, horizontal control networks added new stations, new arcs of triangulation, new measurements for existing arcs, and new rounds of least-squares adjustment to redistribute errors and increase reliability while recalculating geodetic positions. The

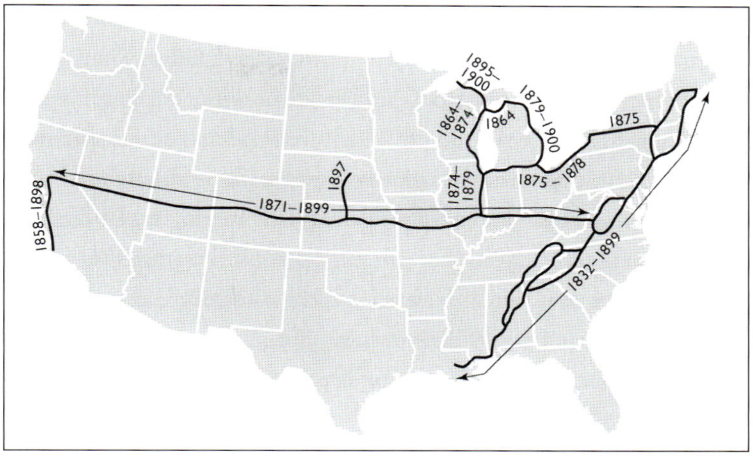

Figure 2.9. Triangulation arcs (with dates of survey) for the United States Standard Datum, created in 1901.

denser horizontal control network led to the North American Datum of 1927, known as NAD27 but not completed until 1932. The central location of Meades Ranch made hand calculations comparatively manageable and reduced the likelihood of a substantial error anywhere on the continent.[63] Hayford had calculated a new ellipsoid in 1909, relying largely on North American triangulation, but even though it afforded a better fit, mapping officials retained the Clarke 1866 figure of the earth for NAD27. A modest reduction in error apparently did not warrant a massive computational makeover.[64]

A different type of survey network established a nationwide series of bench marks, at which a metal disk or other monument marked a specific elevation above sea level.[65] Vertical control networks are anchored to tidal stations along the coast and extended inland by triangulation as well as by leveling along traverse lines. Leveling is an exacting process of moving along the traverse horizontally, step by step with a precision bubble level, while keeping track of the difference in elevation whenever uneven terrain requires a shift to a higher or lower horizontal plane. The National Geodetic Vertical Datum established in 1929, and known as NGVD29, provided a network of bench marks for the elevation contours that gave the US Geological Survey's topographic maps their distinctive appearance.

By the early 1970s, the country and the world needed a new horizontal datum. Denser survey networks had acquired vast amounts of new data, and

technological change had introduced new tools with new benefits and new requirements. Electronic distance measurement was making trilateration an attractive alternative to triangulation, electronic computing made least-squares adjustment faster and more efficient, and artificial satellites and long-range weaponry demanded more exact knowledge of gravity variations and the shape of the planet.[66] Military needs, in particular, suggested that the new ellipsoid should sacrifice improved coverage of North America for a broader but less exact treatment of the Northern Hemisphere. Although the Cold War imperative of targeting Russia with missiles from Nebraska drove the decision, a worldwide ellipsoid was beneficial to peacetime pursuits such as satellite navigation and global climate research.

Work that led to the World Geodetic System of 1984 (WGS84) began in the 1950s, when American military geodesists, concerned with the gravitational pull on long-range guided missiles, called for an ellipsoid that better approximated the geoid than the International Ellipsoid, based on Hayford's work and adopted by the International Association of Geodesy in 1924.[67] Incremental revisions led to an ellipsoid with equatorial and polar radii of 6,378,137 m and 6,356,752 m, respectively, and a flattening of 1/298.257 m—marginally thinner but taller than the Clarke ellipsoid and not as flattened.[68] Unlike Clarke's model, which was intended as a best-fit solution for triangulations on the earth's surface, WGS84 is anchored to the planet's center of mass. As a datum for satellite navigation and GPS surveying equipment, this so-called "geocentric model" did not need to be anchored to the ground at a single point, like Meades Ranch, once an emblematic location for any cartographer who cared about horizontal precision.

WGS84 was a major advance but hardly the last word for scientists to whom millimeters matter. Because the earth wobbles on its axis and the mass beneath the surface is in continual flux, revision continues under the auspices of the International Earth Rotation and Reference Systems Service (IERRSS), a project of the International Astronomical Union and the International Union of Geodesy and Geophysics. In addition to occasionally resetting the planet's clocks by adding a leap second, the IERRSS maintains its own ellipsoid, introduced in 1989 and tweaked in 2003.[69] These adjustments are unlikely, however, to dethrone WGS84 as a stable standard for driving directions and topographic mapping.

Although WGS84 was a welcome advance, its adoption by American mapmakers was slowed by the massive task of calculating new projected positions for millions of point locations as well as for the meridians and parallels that bounded thousands of individual quadrangles. The cartographic bureaucracy's response was the North American Datum of 1983 (NAD83), which was based on the Geodetic Reference System of 1980 (GRS80), a slightly earlier ellipsoid essentially identical to WGS84.[70] Because the new datum relocated the graticule, every topographic quadrangle map had to be reframed as well as revised and reprinted in due course.

Reframing was noticeable but not dramatic. The map excerpt in figure 2.10 illustrates the visual consequences in northwestern Washington State, where the datum shift was greater than elsewhere in the conterminous forty-eight states. Until around 2009, small dashed crosses in the corners of most US Geological Survey (USGS) topographic quadrangle maps cast on NAD27 showed the intersections of the meridians and parallels that would bound the quadrangle under NAD83.[71] A verbal description supplemented the dashed crosses shown here for the lower-left corner of a 1984 map. As the dashed crosses predicted, post-2010 editions of this map, cast on NAD83, ceded to neighboring map sheets small slivers of land along the bottom and left sides but gained compensating slivers from sheets on the top and right—ample evidence of why a new horizontal datum is a massive but rare occurrence.

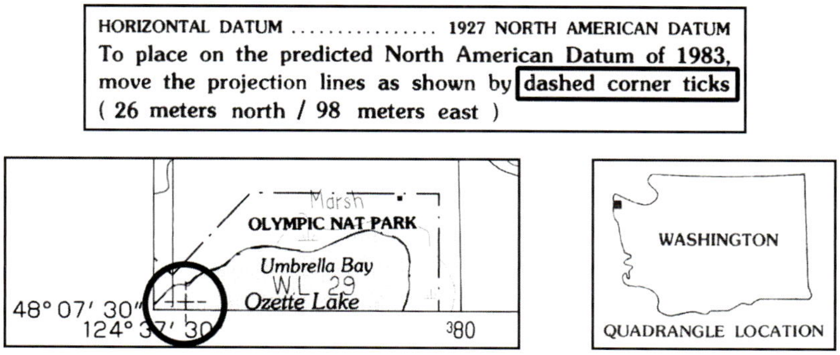

Figure 2.10. Map excerpt from the *lower-left* corner of the 1984 edition of the US Geological Survey's Umbrella Creek, Washington, quadrangle map. Text and quadrangle location excerpts are from the *lower* margin of the map.

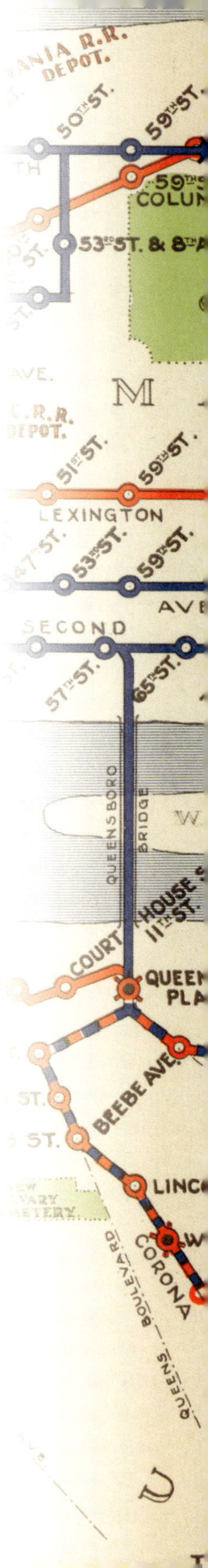

Datum shifts such as this one explain why features that cross adjoining map sheets might not line up, and why people using spatial data compiled at different times, for different areas or different purposes, need to look carefully at the accompanying *metadata* (data about data). Prudence dictates converting everything to a single ellipsoid, typically WGS84. Competent software and computers make it easy to put all features on the same datum.

In addition to anchoring the cartographer's graticule, geocentric world datums—*datums* is the plural of *datum*—also provide a locational framework for satellite geodesy and satellite navigation. Both applications depend on two networks, one on the ground and one in space. Ground stations at fixed, precisely known locations track artificial satellites serving diverse applications, including telecommunications, military intelligence, and the systematic imaging of the atmosphere, the oceans, forests, and terrain in general. Each application has a space component, which might consist of a single satellite or include dozens of satellites, circling the earth in a carefully choreographed network of multiple satellites in multiple orbital planes. That these reference points are in motion makes them no less a network than the ground stations that monitor them.

Satellite geodesy has a distinctive, loosely defined network of satellites, which largely serve other purposes.[72] Tracking stations constantly monitor differences between each satellite's actual altitude and its corresponding intended altitude in a carefully programmed orbit. Discrepancies between observed and predicted positions—geodesists call them *perturbations* or *height anomalies*—have been useful in making detailed maps of Earth's gravitational field as well as in calculating the shape of the geoid, studying the dynamics of the earth's crust, programming flight paths for IBMs, and predicting and compensating for the deflection of a surveyor's plumb bob. Onboard, highly sensitive devices known as *accelerometers* record fluctuations in the downward force known to physics students as the *acceleration due to gravity*.[73] Satellites focused on geodetic work also carry altimeters for measuring height above the ground, as well as specialized ranging instruments for measuring distance to diverse points on the surface, not just tracking stations. Electronic systems for measuring point-to-point distances also allow satellites to track other satellites.

Space-to-space tracking is particularly useful for navigation satellites that follow each other in the same orbital plane. Navstar GPS, the navigation satellite network implemented by the American military between the late 1970s and mid-1990s, includes twenty-four satellites circling the earth twice a day at an altitude of 20,200 km (12,550 mi.) in six orbital planes inclined 55° to the plane of the equator. Each orbit is divided into four slots, and each slot has at least one satellite. Although this "constellation" was designed so that at least four satellites are in view at all times from any point on the ground, it includes additional satellites, some inserted as spares to be repositioned as needed, and others placed in an "expanded" slot to enhance accuracy (figure 2.11).[74] This constantly monitored and carefully managed network relies on space-to-space as well as ground-to-space tracking to identify necessary corrections to a satellite's orbit. Diverse forces such as solar wind and lunar gravity can cause a satellite to lose altitude, and the propellant systems (thrusters) that adjust for these perturbations can also move a satellite into another slot.[75] Satellites that malfunction or are considered no longer useful are decommissioned by boosting them into a "disposal" orbit a thousand kilometers above their operational orbit.[76] Since 1990, a satellite's lifespan has increased from 7.5 to fifteen years, and room must be made for newer, more capable, and more durable satellites.

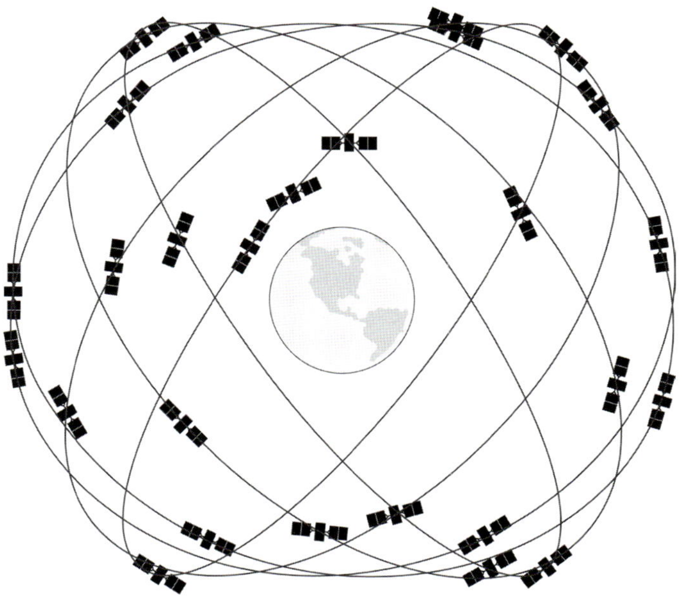

Figure 2.11. Twenty-seven satellites in the "Expanded 24" constellation implemented by the US Air Force in 2010.

Because of precise tracking and carefully programmed orbits, each satellite knows its exact location at all times. And because a satellite can transmit this information to receivers on the ground along with a time signal, any receiver that is "in view" along a straight line, uninterrupted by terrain or structures, can use that time signal to calculate its own instantaneous distance from a known point in space. Because a satellite's radio-frequency signal travels at the speed of light (186,280 miles per second), the microsecond difference between the time the signal leaves the satellite and the time it arrives at the receiver can be used to calculate the intervening distance.[77] That distance describes a hypothetical circle on the ground, which means little because the receiver could be anywhere on the circumference of a circle with a diameter of a thousand miles or more. As shown in figure 2.12, adding a second circle defined by another satellite narrows the receiver's location to two intersection points, and adding a third circle, defined by the time signal from a third satellite, narrows the location to a single intersection, at least in principle. The three circles come close, but they don't intersect exactly at a single point.

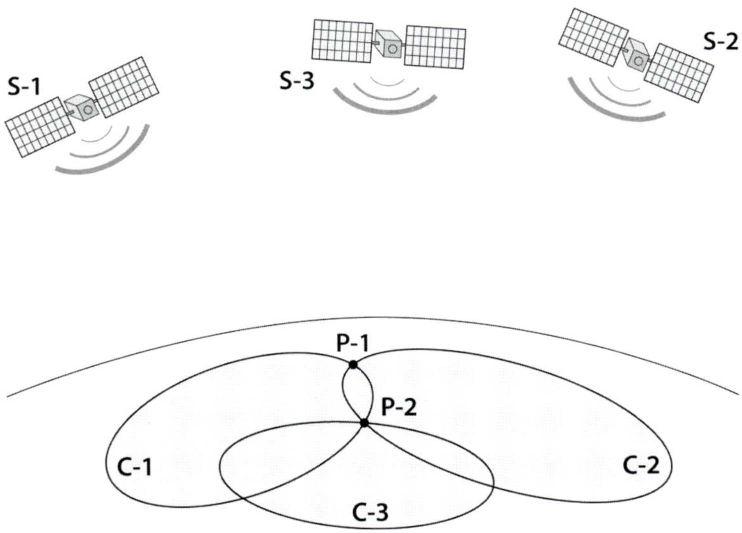

Figure 2.12. Instantaneous circles C-1, C-2, and C-3 defined on the ellipsoid by time signals from satellites S-1, S-2, and S-3 intersect at point P-2. In this example, two circles (C-1 and C-2) define two intersections (P-1 and P-2), which require a third circle (C-3) to remove the ambiguity.

The problem is measurement error, which injects uncertainty into diverse efforts to measure the length of a baseline, the angles of a triangulation network, the elevation of the moon, or the exact time a star crosses the meridian. Uncertainty in satellite positioning reflects the innate difficulty of discerning microscopic differences in time. For each satellite in view, the receiver runs a separate clock that generates and records a pseudo-random sequence of digits at a constant rate.[78] (A small part of this long sequence might look like 01000000 11110101 01011100 00101101.) The satellite's clock simultaneously generates and broadcasts an identical sequence, which the receiver then matches with its own sequence. If the matching algorithm determines that the two sequences are n steps apart and if the clocks added m new digits every second, the time difference between the receiver and the satellite would be n/m seconds. When multiplied by 186,280 miles per second, this time difference defines the radius (in miles) of a sphere centered on the satellite. Any satellite in view will have a similar sphere, each of which intersects the earth's ellipsoid, WGS84, along a circle like those in figure 2.12. Because the clocks' tiny but finite time step 1/m is a source of temporal uncertainty, three circles are not likely to intersect at a single point. In essence, the trio of circles defines three distinct intersections, and the separation of their intersections, however slight, reflects measurement error. Uncertainty is increased by other errors such as small discrepancies between the satellites' atomic clocks, electrical interference in the ionosphere or atmosphere, and multipath signals that bounce off buildings. To get a better fix on the location, including elevation as well as latitude and longitude, the receiver includes the signal from a fourth satellite to calculate an averaged position.

Before May 2, 2000, the military deliberately increased uncertainty by blurring the time signal for civilian-grade GPS receivers, which could be off by more than 100 m, in contrast to military receivers, which were accurate to within 10 m.[79] Rooted in Cold War fears of Russian and Chinese cruise missiles exploiting America's investment, this policy of selective availability was opposed by a consumer electronics industry eager to market vehicle navigation systems as well as put GPS chips in cell phones and tracking devices.[80] Capitalist considerations won out, partly because the Russians and the Chinese were developing their own satellites and partly because the Defense Department could deny coverage to specific regions if a security threat arose.

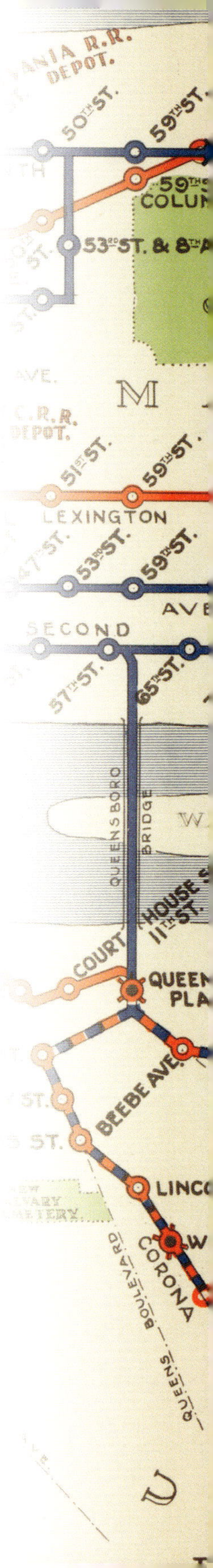

A clever idea known as *differential GPS* helped surveyors and civilian scientists cope with selective availability.[81] This work-around required a second GPS receiver at a precisely known location sufficiently close to receive signals from the same four satellites. Blurred coordinates from the user's GPS were compared with blurred coordinates captured simultaneously at the known location. If the latter coordinates were offset x meters to the east and y meters to the north of the true position, subtracting this offset from the first receiver's readout yielded a markedly more accurate result. Although a receiver's reading could be stored for later correction by software, real-time correction is possible if the two receivers are linked by wireless telephone, shortwave radio, or the internet.

Augmenting a GPS receiver with a single nearby ground control station jump-started a prominent application known as *precision farming*, which uses subfield monitoring of crops and soil to direct carefully measured applications of fertilizer, pesticides, and irrigation water.[82] All that's needed are a detailed electronic map of the farm, a GPS ground station with a radio transmitter, and a tractor equipped with a GPS receiver, a variable-rate fertilizer/pesticide applicator, and sensors for checking soil moisture and crop vigor. An obvious extension was the self-driving tractor, introduced without the fears that undermine public acceptance of self-driving cars.[83]

Wider use of augmented GPS in transportation called for networks of ground stations designed to take advantage of both the lifting of selective availability and major upgrades of the GPS constellation. Without augmentation, a single GPS reading is within 5 m (16 ft.) of the receiver's true position 95 percent of the time, although a high-quality receiver and ideal conditions can shrink the 95 percent radius of uncertainty to less than 2 m (6 ft.).[84] A differential GPS network run by the US Coast Guard uses a network of forty-six remote broadcast sites to give mariners a position error typically in the range of 1–3 m (3.3–10 ft.) along the Atlantic, Gulf and Pacific coasts, the Great Lakes, and navigable sections of the Mississippi River and its tributaries as well as in the Hawaiian Islands.[85] For aviators, the Wide Area Augmentation System (WAAS), anchored by a network of ground stations, promises vertical and horizontal resolution sufficient for automatic collision avoidance.[86]

Integration of two or more networks can be particularly advantageous, as demonstrated in the late nineteenth century when ties between the geodetic triangulation and electric telegraph networks allowed more precise measurements of longitude. In the late twentieth century, the in-vehicle navigation system afforded a different kind of network integration by snapping the satnav's coordinates onto whatever street or highway the vehicle was moving. Although street lines in the receiver's electronic map are also subject to horizontal error, putting the vehicle in the right place on an electronic road map supports not only dynamic route following but also a related speed calculation that is generally more reliable than the vehicle's speedometer, which is vulnerable to tread wear and changes in temperature or tire pressure.[87]

Multiple satellite networks are particularly helpful to the geophysicists, seismologists, and climate scientists who use the International Terrestrial Information Service (ITIS) to track minute movements of the earth's crust as well as sea level rise and the small but measurable effect of atmospheric pressure on elevation.[88] More than 1,700 ground stations worldwide are informed by data from the American GPS network and its two competitors: Russia's Globalnaya Navigazionnaya Sputnikovaya Sistema, or Global Navigation Satellite System (GLONASS), fully deployed in 1996 and enhanced after 2010, and the European Space Agency's global satellite navigation system Galileo, slated to be fully operational in 2020.[89] Two other observation networks—satellite laser ranging (SLR), which measures distance from ground stations to satellites designed to reflect laser light, and very long baseline interferometry (VLBI), which uses radio signals from outer space to measure distances between radio telescopes—give ITIS sufficient precision to calculate the velocity and deformation of tectonic plates.[90]

Drifting continental plates seem not to bother mapping officials in North America, which is separating from Europe by less than an inch a year. By contrast, their counterparts Down Under are wary that the Australian Plate's annual northeastward drift of 7 cm (2.3 in.) is moving the continent's landscape (along with its maps and geospatial data) substantially out of alignment with the graticule used in satellite navigation.[91] This growing disparity is worrisome because of the increased precision of GPS receivers intended to keep driverless cars in their correct lane—who would have thought drifting tectonic plates might someday cause a head-on collision? Rather than overhaul the mapping datum every decade or so, as was done four times in the last fifty years, Geoscience Australia adopted a flexible, two-stage solution. After GDA2020 (Geocentric Datum of Australia

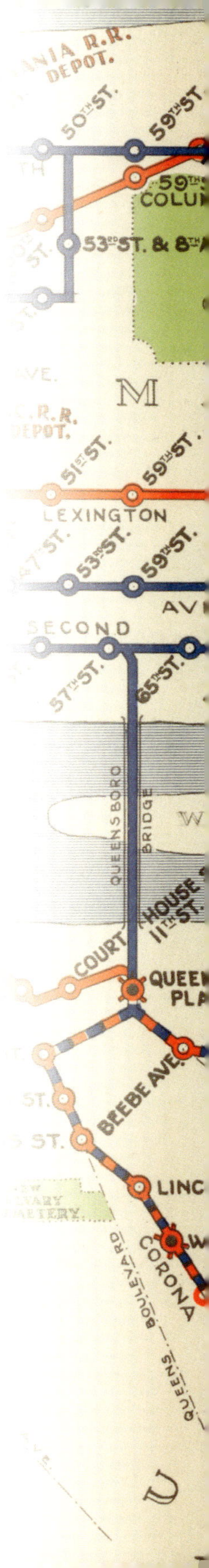

2020) replaces GDA94, from 1994, by shifting point positions roughly 1.8 m to the northeast, software will automatically update electronic map coordinates yearly by whatever small amount is needed to stay current with the satellites' graticule.[92] Moving plates demand dynamic datum transformations if precision coordinates are to keep pace with the landscapes they represent.

3

SYMBOLS

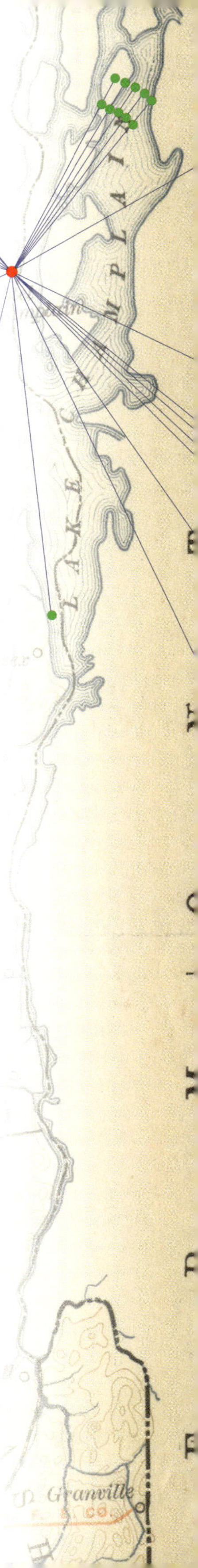

This chapter moves the discussion from measurement and mathematics to representation and meaning. It focuses on the topographic maps developed in the United States in the late nineteenth century, when new kinds of graphic symbols made new kinds of networks meaningful to engineers, investors, public officials, and everyday citizens. After examining the evolving cartographic treatment of a transportation landscape dominated by canals and railways, it concludes with a brief glimpse of how foundational graphic concepts evolved in the latter half of the twentieth century to make airline and urban rapid transit networks meaningful to the traveling public.

Topographic maps depend on geometry as well as graphic symbols, and surveys and graticules provide a geometric framework for mapping networks of all kinds: natural networks such as the streams and rivers that drain the land, informal man-made networks such as paths and wagon trails, and engineered networks such as canals, railways, and highways, with bridges, smooth curves, and aggressively reshaped landscapes. The map symbols that portray networks on the topographic maps encase these functional networks in a systematic, detailed cartographic portrait of a nation partitioned into quadrangles by meridians and parallels—the divide-and-conquer strategy of official mapmakers facing the daunting challenge of producing a general-purpose map series with a uniform scale and consistent criteria for deciding what features to include and how to represent them.

Streams and rivers are a convenient starting point. Almost all topographic maps have them, and they're the simplest mapped network if we ignore complications such as irrigation canals that tap natural waterways, streams that disappear into limestone bedrock, and channels on deltas or alluvial fans that bifurcate rather than merge. Map readers have learned to associate blue lines with flowing water—a convention that conflates wishful thinking about water quality with cartography's need for unambiguous color coding. Although streams with industrial waste or lakes with algal blooms might have a plausible claim to the colors brown or green, established conventions awarded these latter hues to terrain contours and vegetation. However simplistic, reserving blue for water features makes the map's graphic vocabulary easy to learn.

Intuitive coding also helps map readers distinguish between perennial and intermittent streams. Perennial streams, in which water is always flowing (except during times of extreme drought), are represented by continuous blue lines, in contrast to intermittent streams, represented by a pattern of long blue dashes interrupted by three blue dots. Lt. Col. Edwin Roy Stuart (1875–1920), professor of drawing at the US Military Academy, provided black-and-white examples of conventional symbols on official maps in his 1917 textbook *Topographical Drawing* (figure 3.1). To underscore the value of color printing, Stuart noted that watercourse symbols drawn or printed in blue were appropriately "executed in smooth sinuous lines."[1] By contrast, if blue lines are not possible, "a small tremor is forced ... so as to cause irregularity in the lines and thus make a conventional distinction between the lines representing streams and those representing contours."

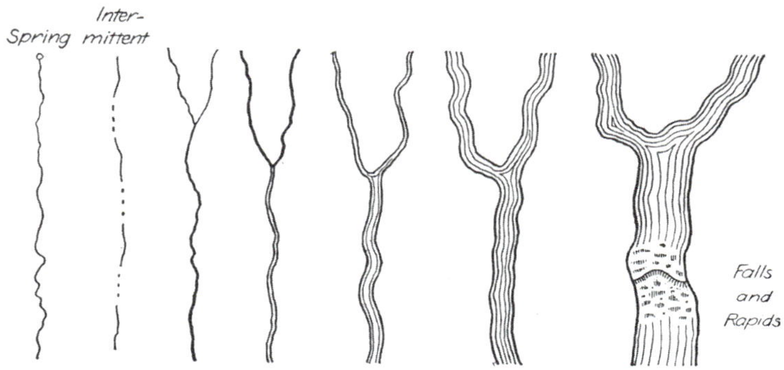

FIG. 29.—Streams.
Conventional color blue. Lines of various weights as shown.

Figure 3.1. Edwin Stuart, professor of drawing at the US Military Academy, included a black-and-white illustration of typical watercourse symbols in his 1917 textbook, *Topographical Drawing.*

The distinction between perennial and intermittent streams is especially important in arid areas, where a continuous blue line assigned to a dry channel might undermine faith in the map's reliability and no line at all would ignore the dangers and origins of flash flooding—an area that didn't receive a drop of rain can suddenly be inundated by drainage from a prolonged heavy downpour more than a mile away. A line's curvature is also important insofar as straight blue lines almost always represent man-made ditches or canals.

Because cartographic symbols must be easily legible, they are often wider than the map scale might suggest. For instance, on USGS topographic maps produced for 7.5-minute quadrangles and published at a scale of 1:24,000, a blue line that's 0.15 mm wide represents a hypothetical corridor 11.8 feet wide on the ground, which is considerably broader than many small streams.[2] And on older topographic maps, framed by 15-minute quadrangles and published at 1:62,500, the same line describes a hypothetical corridor 30.8 feet wide. Because thin blue lines overstate the breadth of a small stream for the sake of legibility, a printed map cannot convey the intricate details of a narrow watercourse, even if an obsessively unrealistic mapmaker wanted to represent all its jogs and meander loops.

Rivers and wider streams pose another problem. When a single blue line would grossly understate a channel's width, the mapmaker delineates the river's banks with lines in medium blue and fills the intervening space with a light-blue tint. On a printed map, a magnifying lens will reveal this tint as a pattern of tiny dots arranged in closely spaced rows, typically 120 to the inch. Before the 1940s, the blue interiors of rivers, ponds, and small lakes were often portrayed by even thinner blue lines running parallel to the banks or shoreline (figure 3.1). For a large body of water such as a bay or the ocean, the thin lines were spaced progressively farther apart until they petered out, perhaps an inch out from the shore. Known as *water lining*, this elegant but tedious technique gave way to so-called "photomechanical processes," which yielded a blue interior after the cartographer or engraver merely delineated the banks or shoreline.[3]

Rules and techniques such as these, and the organizational apparatus to implement them, reflect Congress's decision in 1879 to merge four separate surveys operating west of the Mississippi into the US Geological Survey and three years later to authorize a nationwide geologic "atlas."[4] Because topographic maps would help field geologists plan their explorations and plot their data, USGS

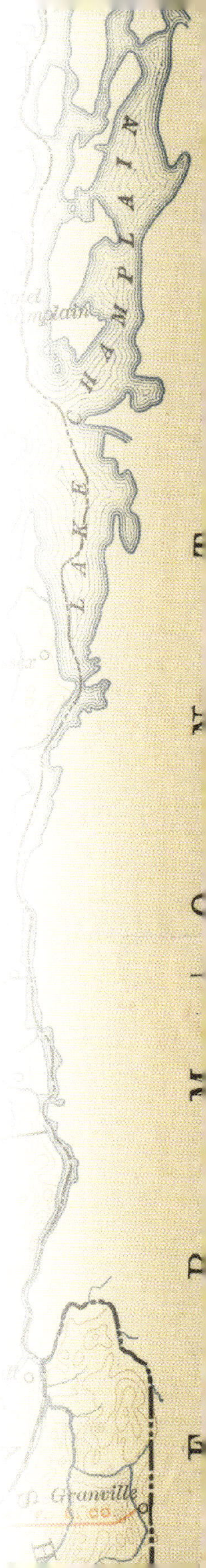

officials saw the geologic atlas as a mandate to produce a topographic atlas as a multipurpose foundation for environmental research, management of federal lands, and economic development, especially in the mining industry.[5] Although the Survey of the Coast, forerunner of the Coast and Geodetic Survey, had been mapping the country's seaward margins since 1832, the Geological Survey was the first attempt to map the interior—all of it—with a systematic topographic survey and uniform map symbols.

Responsibility for planning and overseeing the topographic atlas fell to Henry Gannett (1846–1914), who was appointed chief geographer of the Geological Survey in 1882.[6] The Harvard-educated Washington insider worked for numerous government organizations throughout a four-decade career that included directing geographic operations for the 1880 and 1890 censuses, publishing more than two hundred academic and popular general-interest articles, and leading mapping parties for one of the Geological Survey's predecessor surveys. As chief geographer of the Geological Survey, Gannett established a system of 15-, 30-, and 60-minute quadrangles, mapped respectively at scales of 1:63,360 (later, 1:62,500), 1:125,000, and 1:250,000, whereby an inch on the map represented one, two, and four miles on the ground, respectively.

Gannett outlined his strategy in *A Manual of Topographic Methods*, an 1893 Geological Survey monograph that identified him as the agency's chief topographer.[7] Locally intensive field surveys focused on elevation, slope, and landforms would fill in the gaps between the greater and lesser arcs of the Coast and Geodetic Survey's more geometrically precise network, to which a local topographic survey might not connect directly. When necessary, the topographic survey would create its own primary and secondary networks as well as its own baselines, measured expediently with steel tapes instead of metal bars in leveled boxes on tripods.[8] Though a geodesist might think he was cutting corners, Gannett maintained, "The highest economy is in properly subordinating means to ends," a notion consistent with current thinking among geospatial scientists, who consider map accuracy essentially a matter of fitness for use.[9] Simply put, he noted, "it is not economic to execute triangulation of geodetic refinement for the control of maps upon small scales"—and in his mind, the topographic atlas consisted almost wholly of small-scale maps.

Gannett understood that legible symbols on printed maps lowered the bar for horizontal accuracy. To address the question of how much lower, he suggested, "The maximum allowable error of location may be set at one-hundredth of an inch upon the scale of publication."[10] At 1:62,500, the scale of the most detailed maps the Geological Survey produced at the turn of the last century, a location would almost always be within 50 feet of its correct position. Round numbers such as 50 ft. or 1:62,500 (instead of 1:63,360, exactly an inch to the mile) were consistent with an intuitive assumption that errors would either average out or not noticeably undermine the reliable interpretation of mapped features.

What Gannett meant by "location" was a point with a position and elevation sufficiently reliable for sketching elevation contours and other terrain features but unlikely to appear on the finished map.

> From the constructive point of view, a map is a sketch, corrected by locations. The work of making locations is geometric, that of sketching is artistic. This definition is applicable to all maps, whatever their quality or character. However numerous the locations may be, they form no part of the map itself, but serve only to correct the sketch, while the sketch supplies all the material of the map.[11]

Under this schema, a map's correctness depended on four elements: the geometric reliability of the locations, the number or density of locations, the evenness of their distribution across the map, and "the artistic sense of the topographer, upon his ability to see things in their proper relation, and his facility in transferring his impressions to paper."[12]

Although sketching is artistic work, the topographer's skill set included an understanding of relevant concepts in mathematics, structural geology, and physical geography as well as the ability to apply sound judgment in selecting sketching stations and penciling in contours. Gannett recognized the value of book knowledge supplemented by field training but believed the artistic component, "if not inherited, can be acquired only by long experience in the field, and by many can be acquired only imperfectly."[13] Hire great people, he seemed to be saying, give them whatever apprenticeship seems appropriate, and let them exercise their own judgment. A topographer who can "take a broad as well as a detailed view of

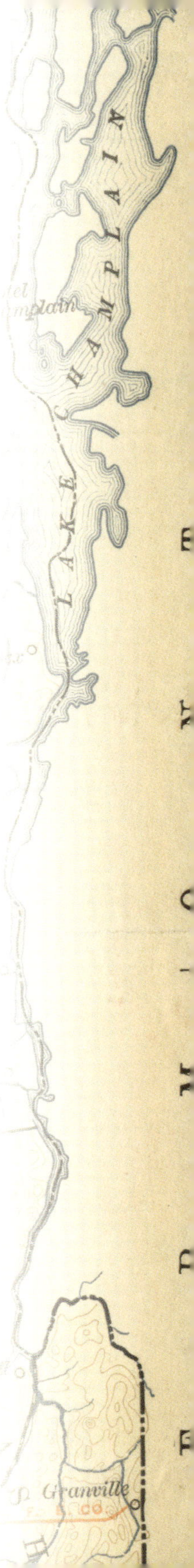

the country and .... understand the meaning of its broad features, and interpret details in the light of those features, ... will be competent to make just generalizations," and know what to include and what to leave out. "It is not possible," he concluded, "to define the degree of detail which the maps should represent."[14]

This thinking explains Gannett's minimal advice on the selection of streams, on how much of their headwaters to include, or even on how wide to make the blue lines—a technical standard apparently left to the engraving department. The closest thing to guidance on drainage is a paragraph highlighting the contribution of "minute rills" [shallow channels] to the larger "network of watercourses."

> Rain falls upon the surface, a portion of it sinks and reappears in springs, while another portion flows down the surface and collects in watercourses, which, joining one another, produce, finally, large streams. During a rainstorm, the entire surface is a network of watercourses, from the most minute rills to the main streams, and in studying transportation and corrasion [by streams], the action of these minute rills, which cover the entire terrain, must be considered as fully as that of the main stream and its primary branches.[15]

The map's scale might limit the amount of detail, but a skilled topographer would strike the right balance and capture the essence of a drainage network.

Though Gannett ignored the final steps of map production, early USGS topographic maps reflect the engraving of three separate printing plates, each with its own ink and distinctive set of features. Engraving and printing were back-office operations, carried out in the Publications Branch, a separate division in the evolving USGS bureaucracy.[16] Topographers would send the engravers three inked drawings, each colored to reflect its unique contribution to the finished map. In *Topographic Methods*, Gannett mentioned color and feature separations only once, in a short section headed "Colors and Conventions," in which he referred fleetingly to the "system of conventions" that governed "work upon the original sheets," on which indelible inks "must not be used."[17] The system was based on three inks: "heavy Prussian blue" for streams, India ink (black) for lettering and "culture," and burnt sienna (reddish brown) for contours and other terrain features. Streams were mentioned only once more, to note the inclusion of "bridges, ferries, fords, and dams upon streams of sufficient importance [and width] to be double-lined," and the remaining two hundred words were divided, more or less equally, between contour lines and the diverse collection of settlements, transport lines, and structures that cartographers refer to as "culture," or occasionally "public culture." The section was dropped altogether when the Geological Survey published an abridged version with the same title in 1906.[18]

Seven years later, the Geological Survey issued a much expanded manual orchestrated by Gannett's successor, Robert Bradford Marshall (1867–1949).[19] Twenty pages with examples of "Conventional Signs," printed in color where appropriate, precede two pages with examples of standardized lettering and a page of acceptable abbreviations. A short preamble notes that the symbols "have been adopted by the United States Geographic Board for use by all mapmaking departments of the government," which accounts for two pages of "Special Military Symbols," such as brigade headquarters, gun batteries, and controlled mines.[20] In 1906, President Theodore Roosevelt had signed an executive order that renamed the United States Board on Geographic Names and expanded its responsibilities for standardizing place and feature names on federal maps to include "the unification and improvement of the scales of maps, of the symbols and conventions used upon them, and of the methods of representing relief."[21] Although President Woodrow Wilson rescinded Roosevelt's order in 1919, eight years earlier a joint federal initiative that included representatives from the Geological Survey, the War Department, and the Coast and Geodetic Survey had produced a comprehensive set of uniform symbols also known as *War Department Document No. 418*.[22] Wary that interagency cooperation might needlessly complicate maps intended for civilian use, the USGS manual warned, "As many of the signs are not used at present by the Geological Survey, none should be used on manuscript sheets unless they also appear on the sheet of conventional signs printed by the Survey."[23]

In addition to illustrating the full range of standardized symbols, the 1913 manual replaced Gannett's pontificating about the importance of sketching, innate artistic ability, and field experience with specific guidance on the selection of features. Tiny streams required a careful balance between completeness and excess. "As a general rule, all perennially flowing streams are to be shown but the map should not be overburdened with insignificant rills and forks, such as abound in well-watered countries."[24] Intermittent streams required a careful distinction between field notes and finished product: between the topographer's "penciled field sheets," on which "all intermittent stream courses are outlined down to their minuter ramifications, as an aid to contour sketching," and the inked compilations sent to the engraving department, which generally excluded intermittent streams "less than three-fourths inch long on the scale of publication." In "more arid districts," however, "the minimum length should be increased up to two inches or more, according to scale, as may seem appropriate to the degree of aridity." In the topographic map's wider context, "drainage lines are delineated not merely because they indicate water features, but because they constitute an important element in the conformation of the land surface and because, especially on contour maps on which lines of greatest declivity are inherently absent, they afford supplementary information of great value in the interpretation of the relief." Because the topographer who correctly interpreted the landscape and its drainage network helped map users make accurate interpretations, he was a graphic journalist of sorts and appropriately identified in the "author diagram" in the lower-left corner of the map sheet (figure 3.2).[25]

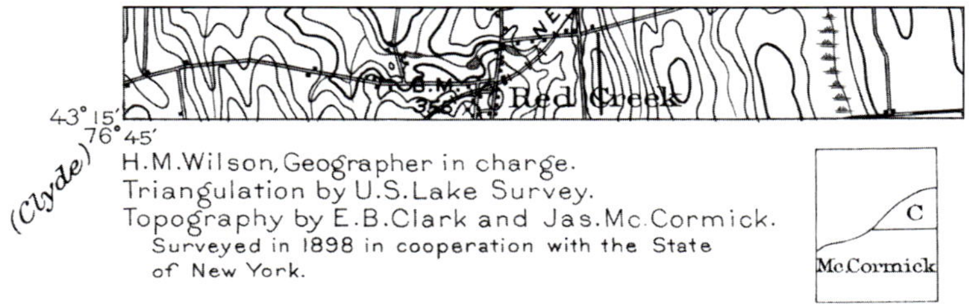

Figure 3.2. The *lower-left* corner of 15-minute USGS quadrangle maps named key contributors to the map sheet. The "author diagram" in this excerpt from the Oswego, New York, sheet, published in 1900, shows that two topographers mapped different parts of the quadrangle.

Few map users, I suspect, are aware of the elaborate behind-the-scenes dance between the drainage network portrayed in blue and the network of drainage divides and ridge lines (or crest lines), which the map user never sees. To understand drainage divides, envision an impervious land surface on which water cannot sink into a thirsty soil. If we ignore evaporation and surface tension, a drop of water falling on this surface should immediately begin moving downhill, pulled by gravity along a path of steepest descent, which in a natural landscape eventually guides it into a stream channel. But there's a line of points from which the drop might move downhill in more than one direction. Except where the drop falls on a local high point such as a rounded dome or sharp peak, it can move in only one of two directions. Called a *drainage divide* because it separates two drainage basins, this line of ambivalence might reflect a prominent ridge crest or be barely discernible to a topographer who walks across a broad, nearly flat upland. And in much the same way that streams merge at a fork, a ridge line can split in two. Because a stream might be intermittent or not merit cartographic representation, in surface geometry its path is called a *valley line*.

Figure 3.3 describes the relationship of elevation contours to valley and ridge lines. Anyone who has studied physical geography or earth science knows (or should quickly recall) that a contour line connects points at the same elevation and that slope is greatest along a path where contours are most closely spaced—that is, along a perpendicular path that descends to the next lowest elevation in the shortest distance. These paths, called *geodesic lines*, always cross contour lines at right angles. Note too that the valley line in the left half of figure 3.3 is marked by a nested set of *V*-shaped contours pointing upstream. Water enters a channel by flowing downslope perpendicular to the contours, and once within the channel, it flows opposite the direction indicated by the *V*s. Experts in map reading know that nested *V*s on a topographic map, regardless of whether they surround a blue line, indicate a channel likely to carry water during and after a prolonged or exceptionally heavy rain. Though my diagram includes only a single valley line, the sequence of *V*s diverging toward the top center could appropriately be enhanced by a blue line representing a tributary. (A curious coincidence: the three-dots-and-a-dash symbol for intermittent streams is Morse code for the letter *V*.)

Ridge lines also cross contours at right angles and can often be recognized by nested contours shaped like a *U* that points downhill. The *U* shape reflects the dominance of weathering, mass wasting, and other processes that attack uplands more uniformly than the erosional processes that cut stream valleys into a landscape and form the sharper *V*-shaped contours. In general, valley lines and ridge lines do not meet, except at points called *saddles* or *passes*.[26]

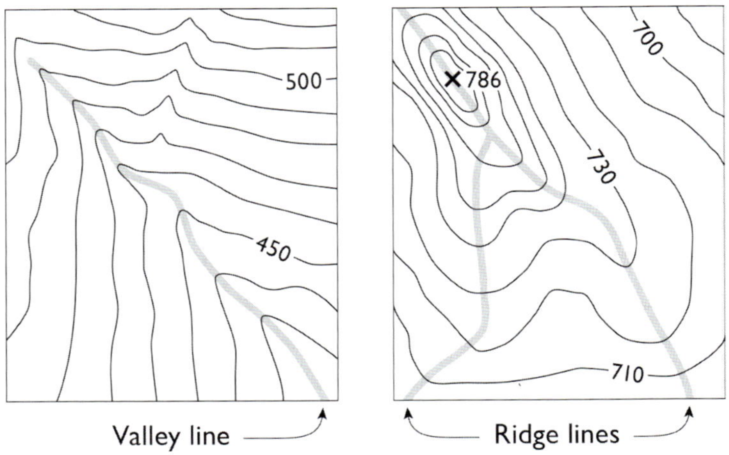

Figure 3.3. Diagrams describing the general relationship of elevation contours to valley lines (*left*) and ridge lines (*right*). Contours with *V*-shaped jogs, in the *left* diagram (*upper center*), suggest a second valley line, perhaps occupied by an intermittent stream in its upper reaches.

USGS topographers used drainage divides to partition a landscape into "constituent masses or unit forms"—a systematic approach that isolates each mountain or hill from neighboring features "by locating the drainage lines that form its natural boundaries, and .... wherever .... practicable, trac[ing] out the crest lines and divides."[27] When added in pencil to the field sheet, these features collectively "constitute[d] the natural skeleton upon which to place the contours." Although map readers never saw the silent network of drainage divides, it was embedded in a warning to topographers not to extend intermittent streams to the ridge crest. This guidance was geometrically explicit insofar as "contours along divides usually suffice to define the locations of stream heads, and the symbol should end between one-eighth and one-fourth inch below these."

Relationships between drainage and contours were also codified in the preparation of prepress artwork. In contrast to Gannett's silence on engraving and printing, the 1913 *Topographic Instructions* explained that hydrographic sheets were inked in Prussian blue, a comparatively dark shade, because features on the topographer's manuscript maps were transferred photographically to the copper plates used by the engraver.[28] Medium or light blue, by contrast, would not give the engraver suitably distinct lines.[29] In drafting their manuscript maps, topographers followed a prescribed "sequence of inking"—"(1) culture, (2) drainage, (3) elevations, (4) contouring, (5) lettering"—which gave man-made features priority and ensured that elevation contours would accommodate, rather than restrict, the drainage network.[30] Where necessary to avoid confusing overlap, drainage lines could be displaced slightly to avoid graphic conflict with culture, and contours were similarly displaced to avoid conflict with culture and hydrography.

The 1913 manual also noted that the images were usually reduced photographically to the scale of the engraved plates (and the printed maps) so that a topographer could draft his manuscript maps at a larger size that allowed a more detailed delineation—a practice that made the engraver an editor responsible for removing details that didn't fit or were otherwise irrelevant. Although the engraver might be a graphic copy editor of sorts, responsible for enforcing the house style, the topographer had authorial control.

Network interactions were particularly prominent in the Geographic Board's standardized map symbols, which not only sanctioned blue, brown, and black lines for drainage, terrain, and culture, respectively, but also included a rich set of graphic codes for networks. One of the four pages with examples of symbols for "Works and Structures" addressed any crossings of relatively wide, double-line rivers and streams that might require a specialized structure.[31] Figure 3.4, which shows the icons for seven different kinds of bridges, omits additional examples provided for fords (3), ferries (2), and dams (1). Short, evenly spaced cross ticks indicate railroad bridges, and pictorial embellishments reflect the engineering design, such as suspension or truss. The three symbols for fords focus on military concerns, such as appropriateness for wagons and artillery, infantry and cavalry, or cavalry alone.

Rules regulated the use of these symbols. Bridges were to be shown for all crossings of double-lined streams, for road crossings of canals and ditches, for drawbridges, and for "all important viaducts over railroads, railroad yards, or streams."[32] In a concession to clarity, bridge symbols were suppressed for commonplace (and quite numerous) railroad bridges and trestles as well as "whenever in centers of dense culture [they would add] to confusion or impair the legibility of the map." Although single blue lines could intersect a road or railway without topographic fanfare—common sense implied a culvert or bridge for railways and almost all roads, and overprinting blue lines and double-line roads did not cause

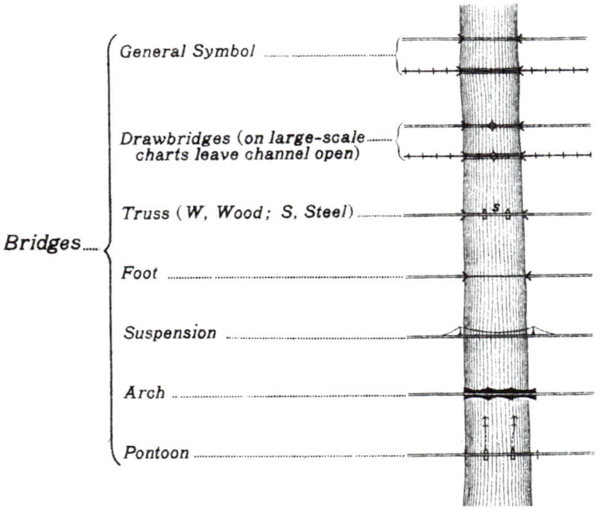

Figure 3.4. Examples in *Topographic Instructions* (1913) included seven symbols showing a double-line river crossed by another network.

visual confusion—a bridge symbol was required when a road crossed a single-lined stream "in sparsely settled regions or wherever the bridge is vital to the use of the road." Although any bridge is arguably essential, a bridge that might necessitate a substantial detour if closed was particularly worth showing.

Canals had a distinctive specialized structure: locks that orchestrated a change in elevation between generally straight sections that crossed few, if any, contours. Figure 3.5 consists of several excerpts from the Syracuse, New York, 15-minute quadrangle map, published in 1898, when the State of New York was still operating an enlarged version of the original Erie Canal. Flanked by towpaths for the mules and horses that pulled canal boats from the Hudson River to Buffalo, this historic waterway would be made obsolete within two decades by the wider and deeper Barge Canal, designed for self-propelled boats.[33] Generally 120 ft. wide, the Barge Canal merited a double-line symbol, whereas the Erie Canal, only 69 feet wide when it passed through Syracuse, is represented by a single line only marginally wider than the 1:62,500 map's single-line streams.[34] I enlarged the excerpts to make key symbols visible, and in converting browns, blues, and some blacks to grayscale, I manipulated the relative darkness of shades of gray to emphasize some features more than others.

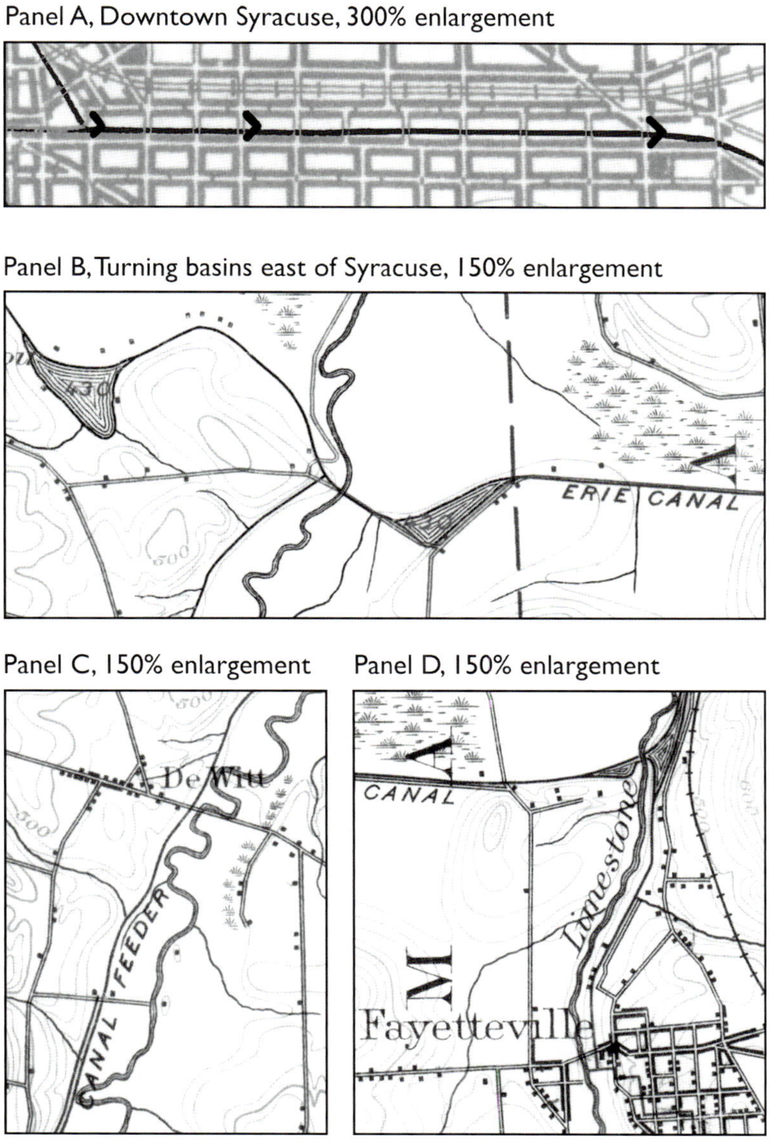

Figure 3.5. Map excerpts, enlarged and converted to grayscale, exemplify the treatment of the Erie Canal on the 1898 USGS topographic map sheet for Syracuse, New York. The short diagonal black line on the *far left* of panel *A* is the Oswego Canal.

A 300 percent enlargement is appropriate for panel *A*, which describes the canal's route through downtown Syracuse, where the dark-gray line (formerly blue) representing the canal competes for space with symbols for buildings, streets, and railroad tracks (now an amalgam of light grays). I emphasized the thick black *V*s pointing up current to mark the locks that lowered westbound boats 26 feet in three steps. Although a single *V* represents one lock, a map at a much larger scale could show a pair of watertight doors at each end of the lock—doors on opposite banks that formed a *V* when closed. When a boat was headed from the upper to the lower side of the lock, a valve was opened to raise the water in the lock so that when the up current doors, or gate, was opened, the boat could move into the lock. The gate was then closed so that when the water was lowered to the level on the down current side of the lock, the boat could move onward after the lower gate was opened. Similar principles allowed a boat to pass through the lock in the other direction.[35]

Enlargement to 150 percent is adequate for the other three panels, in which the map's blue lines are now solid black. Panel *B* was framed to show two so-called "turning basins": wide areas where a boat could change direction. Their triangular shapes reflect sections of the canal converging at almost a right angle. Identical elevation numbers (430 ft.) that interrupt each basin's water lining (originally in blue, now in gray) underscore the canal's inherently level route, diverted to the northeast around a low hill. The broad area devoid of contours down the middle of the panel is the floodplain of Butternut Creek, with meanders characteristic of a natural feature. Note the two black (formerly blue) lines that join the canal from the south between Butternut Creek and the turning basin in the center of the panel.[36] The short straight line is probably a ditch that empties into the canal, and the longer, slightly curved line is a feeder canal delivering water from Butternut Creek as well as from Jamesville Reservoir, 11,706 ft. south of the canal, according to engineering data.[37] This precise length confirms that a navigation canal is a moderately complex hydraulic network that depends on the reliable delivery of water from nearby sources.

In panel *C*, which begins less than a mile south of panel *B*, the label "Canal Feeder" confirms this watercourse's purpose. Note how it captures water from four streams otherwise tributary to Butternut Creek. The Butternut feeder canal was authorized by the Canal Commission in 1872 and completed in 1874. Though history books say the Erie Canal was completed in 1825, decades of subsequent improvements indicate that the work was never finished.[38]

Panel *D*, which overlaps panel *B*—they share the label "Canal"—includes another turning basin, also 430 ft. above sea level. This basin receives water from a feeder flowing northward past the village of Fayetteville in a smooth channel about 200 ft. east of the more sinuous Limestone Creek, which is roughly 10 ft. lower. Built explicitly for the Erie Canal in the 1860s, this feeder delivered water from a reservoir about 15 miles away.[39] Short straight sections connected at right angles follow the alignment of village streets.

About a half mile north of the village, the canal intersects Limestone Creek. A close look will reveal map symbols that suggest the stream crosses above (rather than below) the canal—which seems counterintuitive: if the stream was higher, wouldn't it descend on its own onto the lower ground around the canal? Because the site is only twenty minutes from where I live, I drove over to Old Erie Canal State Historic Park and observed firsthand the mostly intact aqueduct that carried the canal over the stream.[40] This vertical sequence is affirmed by a 1906 decision of the New York State Court of Appeals, which reversed a lower court that awarded damages to a farmer who had sued the state after water from the canal overflowed into Limestone Creek and flooded his field.[41] Intersecting networks can be troublesome legally as well as cartographically.

Intersecting networks can be economically advantageous when a river and canal are headed in the same direction. Canal builders can then avoid the need for feeder canals by dividing a river into level, easily navigable sections separated by dams with locks. And if the canalized river does not extend as far as the intended destination, a conventional canal could complete the route. This strategy is exemplified by the Oswego Canal, which connected Lake Ontario with the Erie Canal at Syracuse and became a key link in a relatively direct all-water route from Canadian ports on the lake's opposite shore to New York City via the Hudson River. Constructed between 1825 and 1828, the Oswego Canal began in Oswego, New York, near the southeastern corner of Lake Ontario, and ascended the Oswego River for 24 miles to the confluence of the Oneida and Seneca Rivers, near the aptly named Three River Point, from which it continued 14 mi. along a nearly straight line into Syracuse.[42] The junction of the southeastern end of the Oswego Canal with the east–west trending Erie Canal is barely visible on the left side of panel A.

The efficacy of a canalized river became apparent when the portion of the Oswego Canal north of Three River Point was incorporated into the New York

State Barge Canal, opened in 1918, and the southern portion into Syracuse was abandoned shortly thereafter. Although canal traffic declined in the latter half of the nineteenth century when an expanding railroad network offered shippers more reliable, year-round transportation, the New York Legislature was committed to providing a low-cost conduit for bulk commodities such as grain and cement. The opening of the Barge Canal also meant replacement of the Erie Canal's original course through downtown Syracuse with a more efficient route about 20 mi. north of the city.[43]

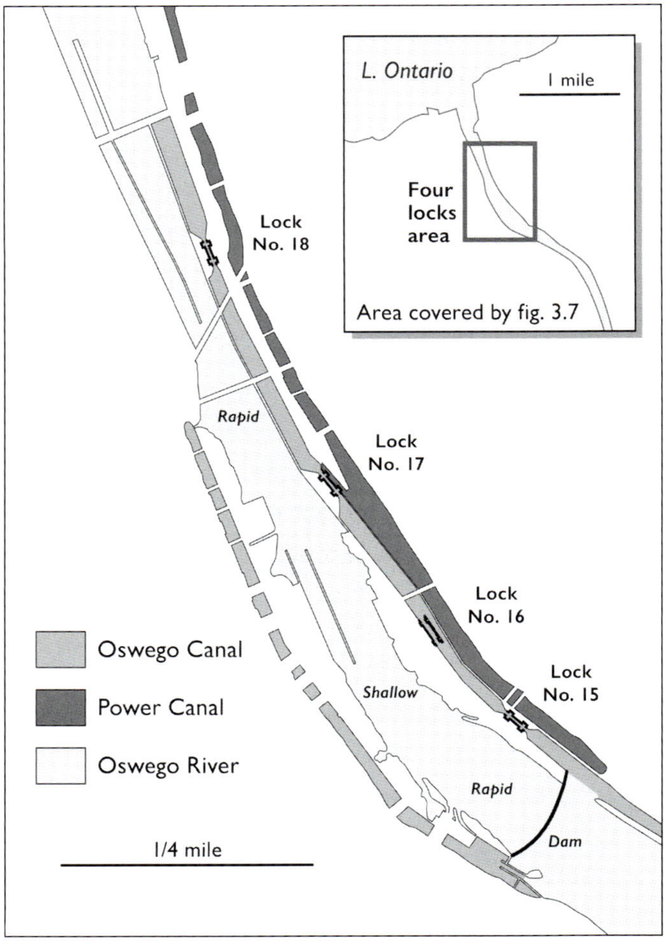

Figure 3.6. Waterways and four lower locks along the canalized Oswego River, as shown on a 1903 Corps of Engineers harbor chart. Interruptions represent bridges.

Amalgamation with the Barge Canal brought a widening and deepening of the Oswego Canal's surviving segment and a reduction in the number of locks within the city of Oswego from four to two. Figure 3.6, compiled from a 1903 Corps of Engineers harbor chart published at 1:8,000 scale, shows the four locks required for the canal's initial ascent of 46 ft. up from lake level, a significant rise underscored by two areas in the adjacent river identified as rapids. Just above the dam, the navigation canal diverges eastward into its hydraulically regulated descent to the lake, and a narrow half-mile stub canal branches off to the west of the dam to serve local shippers. In addition, a power canal served water-powered mills and hydroelectric plants on the east flank of the navigation canal. White interruptions of these canals are bridges or buildings that span the waterway.

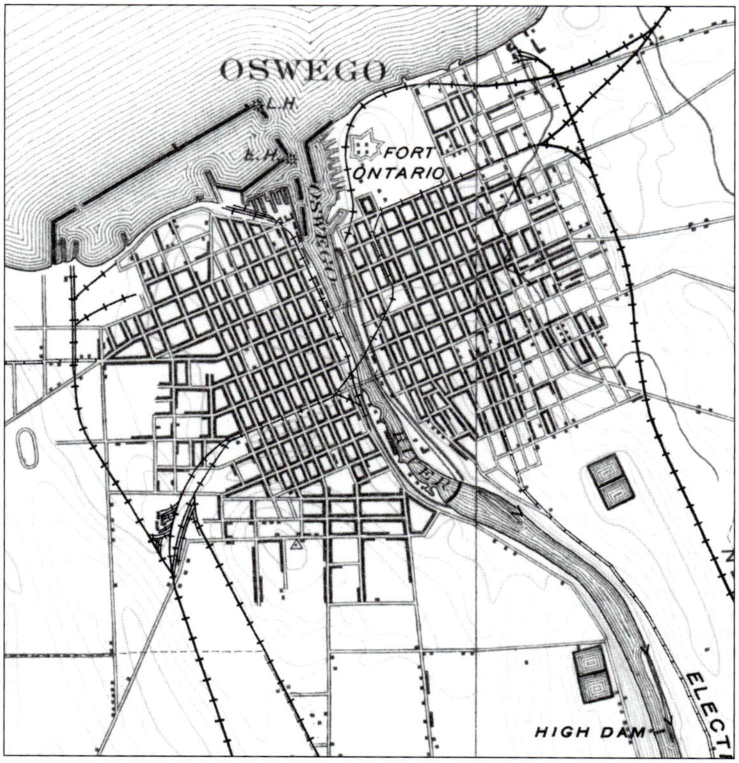

Figure 3.7. Facsimile excerpt from the 1900 Geological Survey's Oswego, New York, topographic map, enlarged to 133 percent. As the locator map inset indicates, the area shown in figure 3.6 is at the *center* of this excerpt.

Oddly, USGS cartographers didn't show any of these four locks on their topographic map of Oswego, published in 1900 at 1:62,500 and highlighted in the enlarged excerpt in figure 3.7.[44] There is no ready explanation for the locks' omission insofar as graphic congestion along the lower canal in Oswego is no greater than in downtown Syracuse (panel *A* in figure 3.5), for which federal mapmakers included three conventional *V*-shaped lock symbols along a fourteen-block section of the Erie Canal, which was crossed by a bridge at every block. That USGS mapmakers were aware of Oswego's canal locks can be inferred from the three conventional *V* symbols in the facsimile, which covers not only the area delineated in figure 3.6 but also a wider area to the south and east. The two *V* symbols near High Dam underscore the map's value to local historians insofar as the High Dam shown in figure 3.7 disappeared a decade later when a new and higher High Dam, constructed about a mile downstream as part of the Barge Canal modernization, submerged the old dam.[45]

Although the foregoing discussion of canal symbols and the detailed treatment of railroad symbols that follows might seem pointless, this level of detail reflects the federal cartographer's commitment to providing a geographic framework relevant to diverse uses, including public works planning and national defense. After all, topographic maps are *basemaps*—that is, basic maps—often used to compile maps for special purposes such as siting factories, electric power lines, and new highways. Invaluable in documenting how an area used to look and function, these symbols constitute an elegant graphical model with an integrity that reflects the network (framework) of the mapped data.

In this vein, figure 3.7 provides a record of other networks prominent in Oswego's history, notably the three steam railroads entering the scene from the south (one to the east of the river and two to the west) and a fourth entering from the northeast. In 1900, Oswego was a vibrant manufacturing center as well as a regional transportation hub prominent not only as the Lake Ontario port served by the Oswego Canal but also as a way point on the New York Central Railroad's line connecting Watertown (to the northeast) with Rochester and Niagara Falls (to the southwest). In addition, Oswego was the northern end of a branch of the Delaware, Lackawanna, and Western Railroad that reached northward 114 mi. from Binghamton through Syracuse. It was also the terminus of the main line of the New York, Ontario, and Western Railway (O&W), which ran 327 mi. south to Weehawken, New Jersey, on the west side of the Hudson River near New York City.[46] Although all three railroad companies have disappeared as corporate entities—rail fans call them "fallen flags"—the rail line east of the river survives as the Oswego Industrial Track, an appendage of the CSX Transportation system serving a recently rejuvenated port facility west of Fort Ontario (a relic of the War of 1812, now a state historical site).[47] The map's rail network is readily apparent in a scattering of company names in less graphically congested areas as well as the distinctive railroad symbol: a thin black line embellished with short, evenly spaced horizontal tick marks. A close look will reveal an interconnected system with a bridge across the river and railroad tentacles skirting the lake shore and river banks.

A close look at the lower-right corner of the map excerpt (figure 3.7) will reveal another kind of railway, a trolley line identified by juxtaposition of the first five letters of "ELECTRIC RAILROAD" with the conventional railroad-in-street symbol, which combines the double-line symbol for city streets or wagon roads with the evenly spaced tick marks of the railroad symbol. In 1900, the Oswego Traction Company ran trolley cars southward five miles along the east side of the river to Seneca Hill and Minetto.[48] A close look at the left side of the map, midway between top and bottom, will reveal a similar symbol marking the trolley line's extension about a mile and a half west from downtown to Oswego Beach, a local amusement park on Lake Ontario. In an era when most Americans depended on public transportation, a recreational destination in the suburbs generated added revenue for the local streetcar company, which usually owned the facility. The streetcar network deserved a place on the topographic map because of its importance to residents as well as its physical imprint on the landscape.

Although early USGS topographic maps can be rich in clues to a city's transportation history, their usefulness depends heavily on when the area was surveyed. Because the maps were cartographic snapshots, fixed in time and seldom revised, they easily missed key additions to, or subtractions from, the local network. For example, the Oswego 15-minute quadrangle map, which was surveyed in 1898 and first published in March 1900, shows the same railroad and streetcar network as reprints dated November 1904, November 1908, June 1913, and 1927, 1932, 1939, and 1948.[49] The only apparent change is the addition between 1900 and 1904 of the label "ELECTRIC RAILROAD"—just the label, not the track

symbol—to the line running out of the city to the west. As a cartographic snapshot, the map fails to capture the decline of Oswego's streetcar service in the 1920s, when automobiles became more widely affordable, as well as a new, more substantial electric line, which lasted only two decades. Although the last streetcar ran in 1927, the trolley company's successor opened an interurban electric railroad down the west side of the river to Syracuse in 1911, too late for USGS topographers. The interurban line was abandoned in 1931.

Enhancement of the railroad-in-street symbol with a descriptive label might reflect a USGS topographer's belated recognition that the intricately engraved cross ticks are barely visible without a magnifying lens—I never noticed the streetcar lines on the framed 1895 USGS Syracuse map decorating a wall at home until I searched for them with a jeweler's loupe. Minute details such as this are characteristic of the array of transportation and communication symbols in figure 3.8, scanned at high resolution from a page at the back of the 1913 *Topographic Instructions* and painstakingly restored in Adobe Photoshop to remove the visual noise of yellowing paper.[50] Reproduced here without enlargement, the page's twenty-five line symbols reflect a clever combination of graphic codes or pictorial metaphors describing the type of movement, the route's capacity or quality, and its technological sophistication. At the top of the page, six symbols, originally printed in blue, differentiate man-made water features such as canals and aqueducts from nonwater networks, printed in black. The next five symbols represent more conventional routes for travel by wagon or on foot, and five more describe different types of railway lines. A map key such as this one gives the curious map reader a scavenger list of hidden gems worth searching for.

First among the four categories of wagon routes are "metaled" roads, so-called because their surface is coated with crushed stone ("road metal" in civil engineering parlance) and presumably graded and drained. The basic parallel-line symbol for merely "good roads" is enhanced by making one of the lines thicker for metaled roads and breaking the pair of continuous parallel lines into dashes—the double-dash line that nowadays typically indicates an unimproved dirt road, suitable for most SUVs and other vehicles with all-wheel drive and better than average ground clearance. Because the Oswego map (figure 3.7) was published in 1900, its delineations do not reveal whether any roads in the city's immediate hinterland were metaled. The sheet of standardized "Lettering

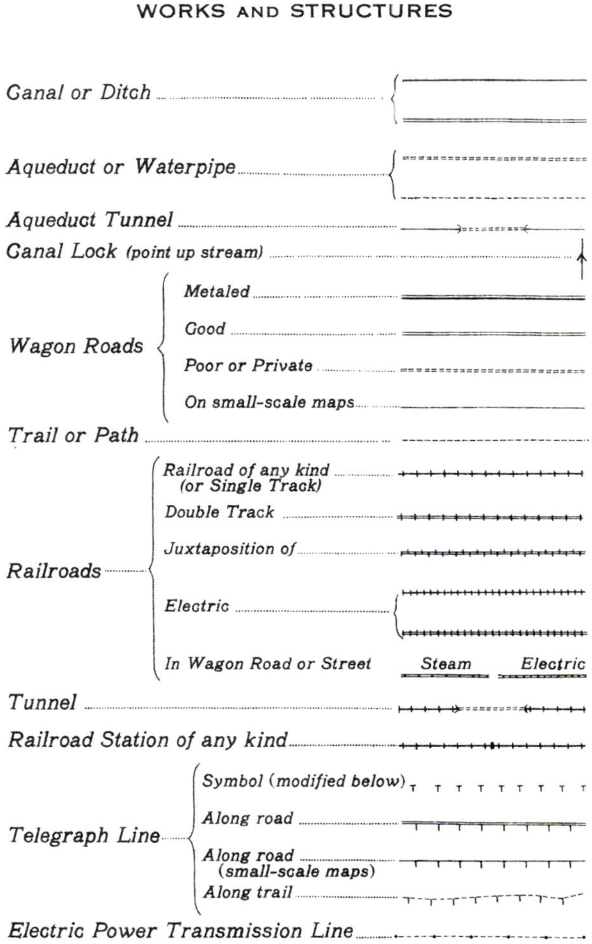

Figure 3.8. Symbolic codes for transport and communications links in the Geological Survey's *Topographic Instructions* (1913). The six line symbols *at the top* were printed in blue.

and Conventional Signs" that the Geological Survey published in 1898 merely distinguished "Wagon Roads," represented by double continuous lines, from "Secondary & Private Roads," shown by double-dash lines.[51] Dashed lines, which suggest a weakened or less certain state, are a long-standing attenuation metaphor also used in cartography to portray uncertainty.[52] Another example is the single-dash lines for footpaths, which remained the same symbol for over a century.

USGS railway symbols changed little between 1898 and 1913. Both symbol sets use double-line and single-line tick-marked symbols (figure 3.8) to differentiate double-track from single-track railroad lines, a distinction important because the former could accommodate two-way traffic more efficiently, without the need for passing sidings. The two symbol sets also used the spacing of tick marks to differentiate steam and electric railroads: for electric railroads, typically constructed with lighter-weight rails and mostly limited to passenger service, the tick marks were noticeably closer than for steam railways, which carried most inland freight shipments over more robust tracks. More closely spaced ticks also differentiated an electric railway "in [a] Wagon Road or Street" from a steam railroad sharing a city street with vehicular traffic. Look closely at the Oswego map (see figure 3.7), and you will find an example south of Fort Ontario.

Not every available railroad symbol was required for the Oswego topographic sheet. For instance, nowhere will you find the official symbol for the "Juxtaposition of" separate railroad lines running side by side (figure 3.8). Although Lackawanna and New York Central tracks were juxtaposed for several blocks just west of the single railroad bridge across the river, the USGS topographer preferred a single railroad-in-street symbol (see figure 3.7), which seems reliable insofar as a narrow roadway might have separated the two sets of tracks at one time.[53] Although the juxtaposed-tracks symbol was seldom required, instances abound on the 1901 15-minute quadrangle map of Chicago, the country's quintessential railroad hub.[54]

The Oswego map also omits the railroad tunnel symbol (figure 3.8), even though the tracks went underground for a short stretch about two blocks after crossing to the east side of the river, just before they turned northward. Abandoned when the O&W went bankrupt in 1967, the train tunnel is now part of the half-mile O&W Railroad Pedestrian Promenade and Bikeway.[55] It is clearly labeled on the 1903 Corps of Engineers harbor chart that provided the canal and locks delineations for figure 3.6.

Federal mapmakers did not always follow prescribed graphic codes. Close inspection will reveal that the USGS topographer stopped using the streetcar symbol once the line entered the built-up area: an omission confirmed by other maps of the Oswego trolley network as well as by common sense—most inbound streetcar riders were headed downtown. By contrast, the 15-minute Syracuse quadrangle, published in 1898, appears to show all trolley lines extant when the city was surveyed, in 1895.

Also apparent is the topographer's neglect of the small buildinglike square prescribed for a "Railroad Station of any kind" (figure 3.8). Oswego was served by three rail lines, each with its own passenger station, but none is represented by the "conventional symbol." Still, this neglect is hardly unique to the Oswego sheet. Although the 1913 *Instructions* codified the symbol's usage—"A railroad-station building should be represented conventionally by a large or elongated house symbol placed across the tracks."[56]—I cannot recall ever seeing a USGS map that used the prescribed station-astride-the-track symbol. By contrast, this symbol is common (but hardly universal) on European topographic maps. For example, British topographic maps at a similar scale typically identify railway stations by either welding a square or circle onto the track symbol or adding "Sta." to a track-side building. Both approaches became common practice across the Atlantic, where the passenger rail network was denser and arguably more important than in North America.[57]

Railroads posed another challenge to topographers, who had to determine not only whether a siding or spur could be included but whether it was worth including. The *Topographic Instructions* issued in 1913 and 1918 offered the same limited guidance—"Double tracks, railroad yards, spur tracks, and switches should be shown so far as the scale will permit"[58]—and the slightly more expansive advice issued in 1928 focused on inking and engraving.

> In railroad yards, parallel spur tracks, etc., only as many tracks should be inked as can be legibly engraved on the publication scale, as too many tracks make difficult inking, illegible advance sheets, and impracticable engraving. Where switches and sidings are inked alongside single tracks, both the main track and the side tracks should be inked in finer lines than the main track elsewhere; and these fine lines should be inked first and the extension of the main track inked afterward in a heavier line, to make clearer copy for engraving.[59]

In short, USGS topographers had at least as much discretion in selecting the railway network's sidings and spurs as they had in delineating the drainage network's headwaters.

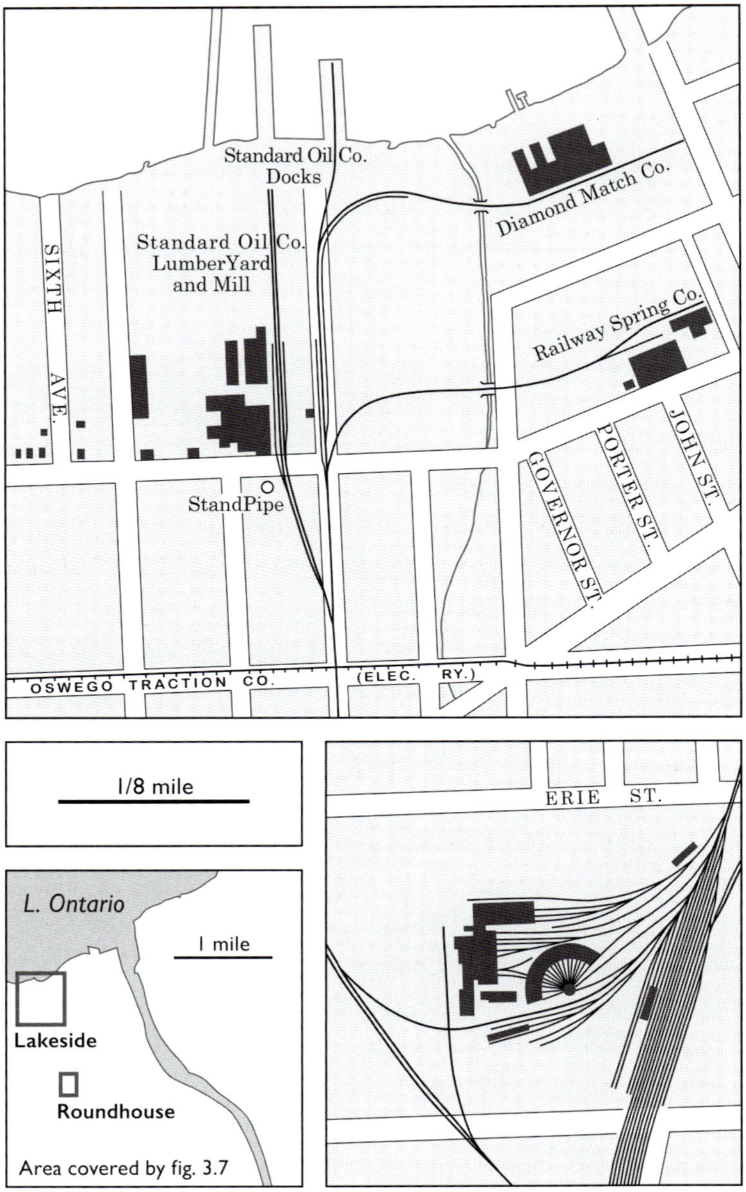

Figure 3.9. Corps of Engineers 1:8,000 harbor chart (1903) shows network details that had to be generalized when compressed into markedly smaller spaces on the 1:62,500 topographic map in figure 3.7.

How freely Oswego's topographers[60] exercised their cartographic license can be inferred from the network features on the two maps in figure 3.9, copied carefully from the comparatively detailed 1:8,000-scale 1903 Corps of Engineers harbor chart I used for figure 3.6.[61] I call these neighborhoods Lakeside and Roundhouse because the top map in figure 3.9 shows multiple railroad spurs approaching Lake Ontario from due south, and the second is a detailed view of the area around a New York Central roundhouse with a turntable and thirty stalls.[62] The Corps of Engineers map appears to include every switch and every siding in both neighborhoods.

Comparison of figures 3.7 and 3.9 indicates that USGS topographers captured key features of the Lakeside railroad network—namely, the two straight north–south tracks perpendicular to the lake and the two separate tracks that diverge to the east to serve facilities that the harbor chart identifies as the Diamond Match Co. and the Railway Spring Co. Because of limited space, neither factory building appears on the USGS map (figure 3.7), which also ignores the extension of one of the north–south spurs onto a Standard Oil Co. dock. Less important details that had to be suppressed include the half-dozen sidings that parallel the tracks shown, the gentle eastward curvature of the spur into the match factory, and the trolley line's continuation across the bottom of the map. Note that the 1:8,000 Corps of Engineers map in figure 3.9 not only identifies the "Oswego Traction Co. (Elec. Ry)" by name and motive power but also describes the shift of the single-track trolley line to the edge of the roadway once outside the built-up area. Side-of-the-road operation, with only a single set of poles, was less costly than a trolley wire suspended over the center of the street.

Graphic compression is markedly more severe at the Roundhouse site (*lower right* of figure 3.9), for which the smaller-scale USGS map cannot retain the shape of either the roundhouse or its turntable. Although the generalized map in figure 3.7 necessarily sacrifices the rich details of the railyard and service facility—the yard's dozen-plus parallel tracks shrink to three—it preserves key trends, such as separate tracks entering the area from the southeast and the northwest, the two separate lines diverging to the northeast, and the triangular junction of three tracks, which allows a locomotive to reverse direction in a maneuver akin to making a three-point turn with an automobile. Although a railroad's turntables, roundhouses, sidings, spurs, and rail yards are functionally essential parts of the network, their geometric details are analogous to capillaries, forced from view when a reduction in map scale dictates a focus on arteries.

Among cartography's quintessential trade-offs is the wider geographic scope possible when map scale shrinks—in web cartography, we take advantage of this trade-off whenever we zoom out. In the world of paper maps, military strategists and transportation planners recognized the need for a broad regional or megalopolitan overview that shows not only transport arteries but also the terrain that channels freight, passengers, troops, and surface runoff along a limited number of well-defined paths. In the years after World War II, Defense Department mapmakers met this need with a nationwide series of 1:250,000-scale topographic maps compiled from published topographic maps as well as aerial photography, which was particularly valuable where existing coverage was lacking or out of date. In 1958, the Geological Survey was given responsibility for the 1:250,000 series, which at the time was the only map series that covered the entire country.[63]

Figure 3.10, an excerpt covering the area south of Lake Ontario near Oswego, illustrates the 1:250,000 maps' transport focus and wider geographic scope. To promote comparison with the rail network on 1:8,000 and 1:62,500 maps, I used Photoshop to mollify the otherwise muddled grayscale conversion of the map's strong greens (wooded area), yellows (urbanized area), blues (water features), and reds (major highways), not to mention its visually weaker but more numerous brown elevation contours. Although mainline tracks remain, gone are the industrial spurs and generalized railyards visible on the 1:62,500 15-minute topographic map but wholly irrelevant to strategic regional planning. Even so, generalization of the railroad network is no less harsh than this smaller-scale map's simplification of highways, drainage, and terrain. Moreover, a few enhancements are apparent: pictorial symbols pinpointing airports and landing fields, route numbers for key state and federal highways, and (if you look carefully) numbers indicating distance in miles between red stars placed along major roads. In addition, a thickening of the railroad symbol sometime before 1965 made the generalized rail net more visually prominent.[64] In the late 1920s, Army Corps of Engineers aeronautical charts used thick lines like these to emphasize the "iron compass" used by pilots for visual navigation.

Comparison with the map's earlier (1948) edition would confirm that the line that once ran east from Oswego toward Watertown was cut back to

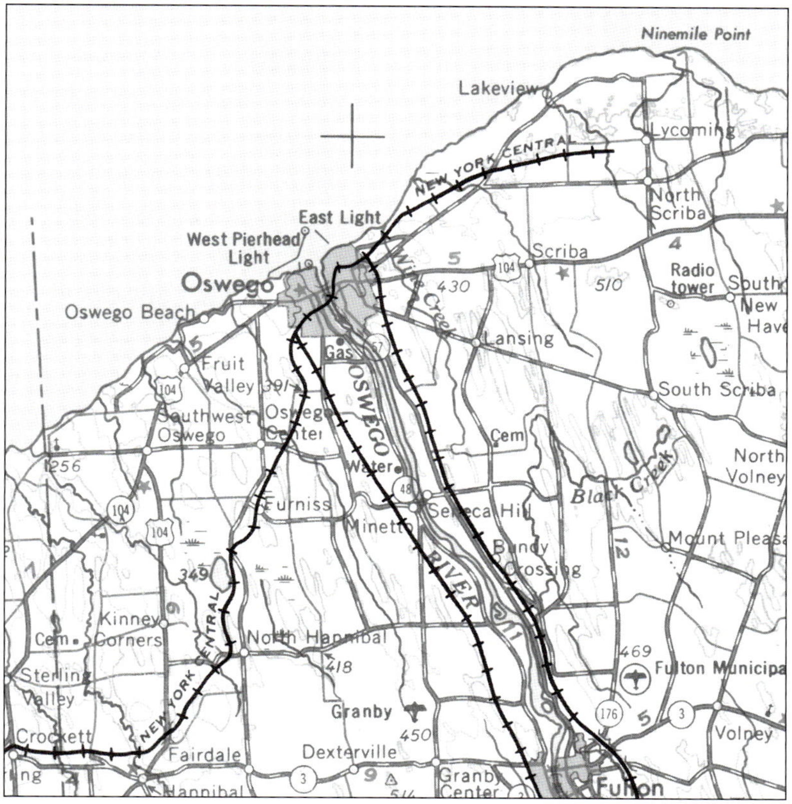

Figure 3.10. Oswego is on the Rochester, New York, sheet of the 1:250,000 strategic planning map series initiated in the late 1940s by the Army Map Service. The series was transferred to the Geological Survey, which updated the Rochester map in 1965.

Lycoming, where it connected with a spur (not shown) to a trio of nuclear power plants under construction at Nine Mile Point. More recent editions would reveal further pruning on the west side of the Oswego River. The track to the southwest to just beyond the edge of the map has been removed, and the other line, running west of the river toward Fulton, is still in place but has been out of service for several years. Networks are dynamic features, but since the early twentieth century, the nation's railroad network has lost more track than it gained.

Government transportation officials concerned with rail net contraction recognize two levels of link severance. More severe and almost always permanent is *dismantlement*, defined as the removal of rails, ties, and salvageable structures such as bridges. By contrast, an *abandoned* railroad, though vulnerable to rust, termites, and invasive vegetation, is otherwise complete. "With ballast, track, and bridges substantially intact," the Geological Survey observes, "it could be put into limited use with a reasonable amount of repair work."[65] This distinction is exemplified in facsimile map excerpts for two areas near Syracuse (figure 3.11). The upper trio of maps depicts a branch line's deterioration from operating, *left*, to abandoned, *center*, to dismantled, *right*. The word "abandoned" reinforces the graphic attenuation of the conventional railroad symbol—cross ticks on a long-dash line—whereas appending "old railroad grade" to the short-dash trail symbol indicates that, with rails and ties gone, the lineation is a high-and-dry trail that was once a railroad. In the lower panel, parallel contours that bracket the former rail line underscore its slightly elevated position, which sustains a trail (single dashed line) or narrow road (double-dash line) across the ballast left behind. The straight right-of-way is also a convenient corridor for the powerline portrayed by the dot-and-five-dashes symbol—might not the dots suggest poles or towers?

Adding "abandoned" to the attenuated, dashed-line-with-cross-ticks symbol reflects a thoroughness accorded other networks portrayed on topographic sheets published without a map key or legend. Dashed lines labeled "pipeline" or "telephone" do not require an explanation, after all, and even though a novice map user might readily associate a cross-ticked line with a railroad, the operating company is typically identified somewhere along the symbol's extent. Replacing an abandoned link's corporate identity with "abandoned" is surely more reasonable than the arcane symbol recommended by the Federal Board of Surveys and Maps, which in 1938 published the booklet *Suggested Symbols for Plans, Maps, and Charts*.[66] As shown in figure 3.12, the recommendations endorsed the USGS treatment of intact railways (see figure 3.8) but proposed the puzzling addition of tiny check marks or freehand *V*s to designate an abandoned rail line. Though the symbol is undeniably distinct, its graphic logic is (to be kind) obscure. I have never seen it used on a published map or in any manual.

In 1957, abandonment claimed one of Oswego's three railroads: the 541 mi. O&W, the first Class I American railroad to be completely taken out of service, sold off, and mostly dismantled.[67] This unprecedented corporate collapse is not evident on the 1:250,000 topographic map published in 1965 (see figure 3.10) because the tracks between Oswego and Fulton were absorbed by the New York

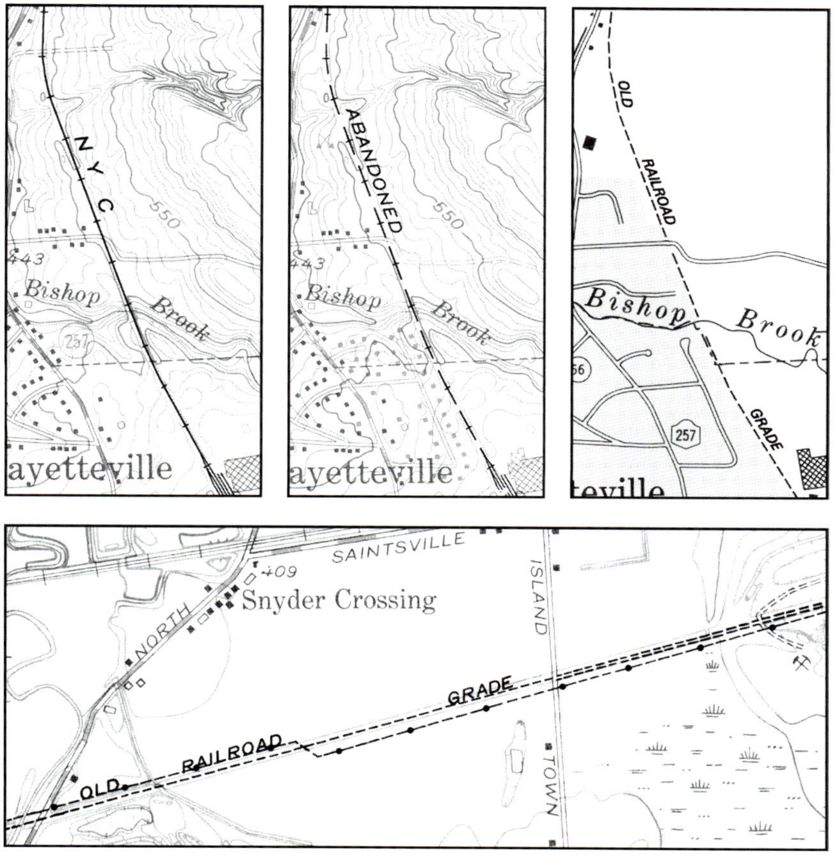

Figure 3.11. Cartographic snapshots at 1:24,000 illustrate abandoned and dismantled railroads. *Top*: The progressive decline of a New York Central Railroad branch north of Fayetteville, New York, described as operating in 1957, abandoned in 1978, and dismantled in 1990. *Bottom*: A trail (dashed line) and unimproved road (double-dashed line) repurposed the straight, level, well-drained roadbed of a dismantled railroad east of Syracuse. Excerpted from 1957 and 1978 USGS maps of East Syracuse, 1990 New York State Department of Transportation map of Syracuse East, and 1957 USGS map of Manlius, New York.

Central Railroad, and remain in operation as part of the CSX system.[68] Although tracks serving shippers in cities with another railroad often survived, the dismantled O&W left scars on the landscape that persist on maps as suggestive contour lines or more explicitly as an "old railroad grade."

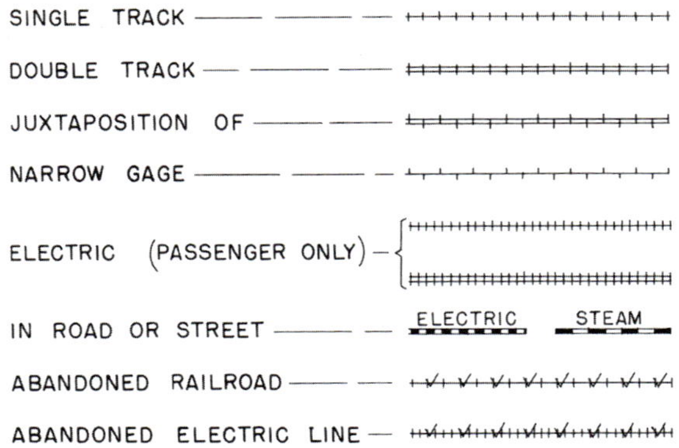

Figure 3.12. US National Resources Committee, *Suggested Symbols for Plans, Maps, and Charts* (Washington, DC, June 1938), plate: "Transportation: Railroads, Bus Lines, Subways, Pipe Lines, Etc.," plate F-2.

The O&W's demise was the culmination of decades of declining traffic, labor strife, and bankruptcy litigation. Although its official map (figure 3.13) shows a comparatively direct mainline between Lake Ontario and New York City, this smooth cartographic backbone disguises a devious path connecting communities willing to buy railroad bonds because they feared the consequences of being bypassed by a major carrier such as the New York Central or the Delaware Lackawanna.[69] A picturesque route through the Catskills generated tourist traffic but added to operating costs, and the belated addition of branches such as the double-track connection to Scranton, Pennsylvania, failed to generate freight traffic sufficient to keep the system solvent.

Railroad maps such as figure 3.13 are a distinctive cartographic genre that combines cartographic license with the hyperbole of advertising and marketing.[70] Smoothed or straightened routes suggest geographic efficiency, competing carriers are ignored or disparaged, and purposeful stretching creates space for copious place-names. This abundance of potential destinations reflects the mapmaker's reliance on wax engraving, a reproduction technique whereby names constructed letter by letter on a straight or slightly curved composing stick were stamped into a thin wax coating applied to a metal plate. This plate became a letterpress printing plate after electroplating added metal where stamping had exposed the metal.[71] Before stamping names on the plate, the engraver inscribed route lines and boundaries with a sharp or blunt-point engraving tool called a *burin*.

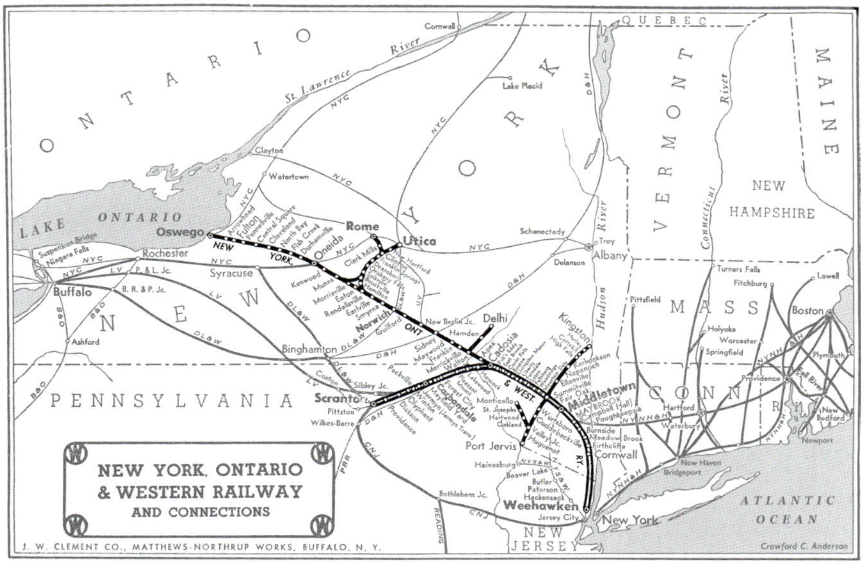

Figure 3.13. Wax-engraved map of the O&W was used in timetables and directories. From *Official Guide of the Railways and Steam Navigation Lines of the United States*, June 1957.

Before the engraver inscribed the lines and labels, a clever cartographer created a customized projection such as the O&W map, which stretches central and eastern New York State, decreases the distance between Oswego and Buffalo, and narrows New England and New Jersey. According to map historian David Woodward, who wrote the definitive history of wax engraving, the Matthews-Northrup Works, which produced the O&W map, had "one of the finest map-engraving departments in the United States."[72] Although map drafters typically lacked formal training in cartography, they were supervised by "chief cartographers with at least some geographical background."[73]

Clever distortions are particularly prominent in maps of urban transit systems, the exemplar of which is the *London Underground Map*, conceived in 1931 by Henry Charles Beck (1902–74), a draftsman employed by the London system.[74] Modified several times by London Transport, Beck's design is reflected in the maps of most transit networks, with multiple lines connecting named stations spaced farther apart outside than inside the central city.[75] Typically, each route is shown by one or more horizontal, vertical, or diagonal sections with more or less evenly spaced station symbols embedded in the line symbol. Cartographers recognize this as a type of *linear cartogram* focused on connections and legibility.[76] Although my example (figure 3.14) uses shades of gray, each route is usually represented by a distinctive color—Washington (DC)'s Metro even named its routes for their map colors. Usually a scattering of surface features is added as a frame of reference, and the map's geometry is manipulated to allow more detail in the downtown area, where lines converge. Large, poster-size maps are displayed in stations, and much smaller versions fit conveniently in a shirt or coat pocket.

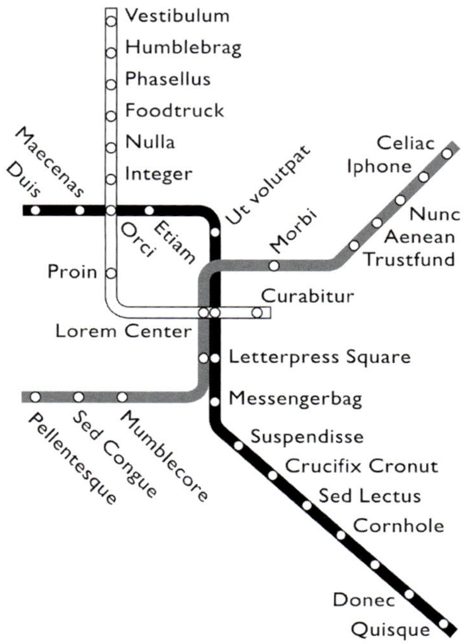

Figure 3.14. Typical design of linear cartogram for an urban rapid transit system.

Another prominent network map focuses on the hub-and-spoke route structure described by the hypothetical example in figure 3.15. As the term implies, the typical destination is connected to a single hub, connected in turn to the network's other hubs. This configuration followed the deregulation of United States airlines in the late 1970s, which led to a consolidation of the industry into fewer, larger companies.[77] Unlike urban subway maps, designed to promote travel within a city's only available rapid transit system, airline maps are more an advertising tool than a trip planning device. Although the airline connections described graphically in maps such as figure 3.15 are embedded in the electronic databases that inform a traveler's online search for a convenient or economical itinerary, the maps are largely advertising devices designed to burnish the company's public image. In this sense, many network maps play a largely rhetorical role distinct from network maps that support spatial analysis or a historical narrative.

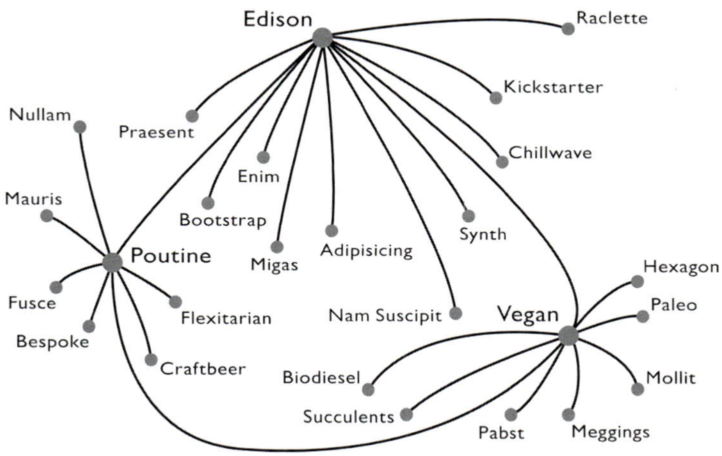

Figure 3.15. Typical design of a hub-and-spoke airline map.

4

INFRASTRUCTURE

Networks such as roads and power lines connect with cartography in multiple ways. Topographic maps depict them as landscape features, and thematic maps highlight their connections and service areas. And because maps are essential for locating routes, acquiring property and easements, planning excavation and drainage, designing and constructing permanent structures such as bridges and towers, and navigating the often sinuous path through or around environmental regulations, there would be no rights-of-way without maps. In addition, underground facilities such as water, gas, and sewer lines, as well as buried electrical and telecommunications cables, depend on cartography for adding customers, repairing breaks, and avoiding disruption by unaware excavators. This chapter addresses the historical role of maps in planning and managing infrastructure, including canals, railways, highways, power lines, and underground utilities. It begins with a look at the Erie Canal, the massive government-run infrastructure project acclaimed as the country's "first school of engineering."

Canals were a marvelous invention that extended inland the fluid movement of cargo ships pushed by winds and currents across level oceans. Long, low canal boats pulled by draft animals on an adjacent towpath could move much heavier loads than nineteenth-century freight wagons dragged by horses and mules along roads that offered little reprieve from gravity or friction. Historians estimate that a canal boat with forty tons of coal could advance steadily behind a single horse, whereas a four-ton wagon on a road with ruts and unavoidable undulations required two horses.[1] A canal system achieves this efficiency with a network of reservoirs and feeder canals that leverage the hydrologic cycle, powered by the sun,

which drives evaporation, creates winds, and delivers rain from distant oceans. Gravity, in turn, funnels water into locks so that boats moving upstream can rise smoothly to the next level and drains the same locks so that boats headed downstream can descend gracefully, without the frictional resistance of the brakes that slow wagons on steep grades.

Europe's canals had demonstrated their efficiency in moving bulk commodities long before 1803, when New York's surveyor-general Simeon DeWitt (1756–1834) and US senator Gouverneur Morris (1752–1816) pondered the prospects of a canal connecting the growing western and central portions of the state with New York City.[2] The most obvious route was the natural corridor from the Hudson River near Albany to Lake Erie near Buffalo (figure 4.1). This route includes the Mohawk River, which joins the Hudson several miles north of Albany and runs westward about 125 miles to a portage dubbed the "Great Carry," near Rome on the drainage divide between the Hudson and the St. Lawrence River. Native traders would lift their canoes out of the water and carry them six miles overland to Wood Creek, which flows into Oneida Lake, connected to Lake Ontario by the Oneida and Oswego Rivers. In 1728, European merchants adopted this ancient 200-mi. route by establishing a trading post at what became the city of Oswego.[3] Although a canal tying the Mohawk River to Lake Ontario near Oswego could be built less expensively than a longer canal across western New York, an interior route through the state's agricultural heartland would avoid the lake's whimsical and occasionally treacherous storms. Morris saw the steady, predictable movement of canal boats pulled by horses on an adjacent towpath as a clear advantage.

Engineers were certain a canal was feasible if the money could be found. Private investors had already built three American canals, but none was longer than 30 miles.[4] No consortium of capitalists was willing to undertake the massive effort of connecting the Atlantic Ocean with Lake Erie, and the federal government had little interest in canals. In 1806, Congress had authorized the Cumberland Road as a strategy for connecting the mid-Atlantic states with the trans-Appalachian West. Regional politics were, no doubt, involved. In the 1820s, Congress refused to support the Erie Canal, and in 1830, when the House of Representatives was asked to approve the Washington National Road Bill, seventeen of New York's twenty-two House members voted no, despite strong support elsewhere in the Northeast.[5]

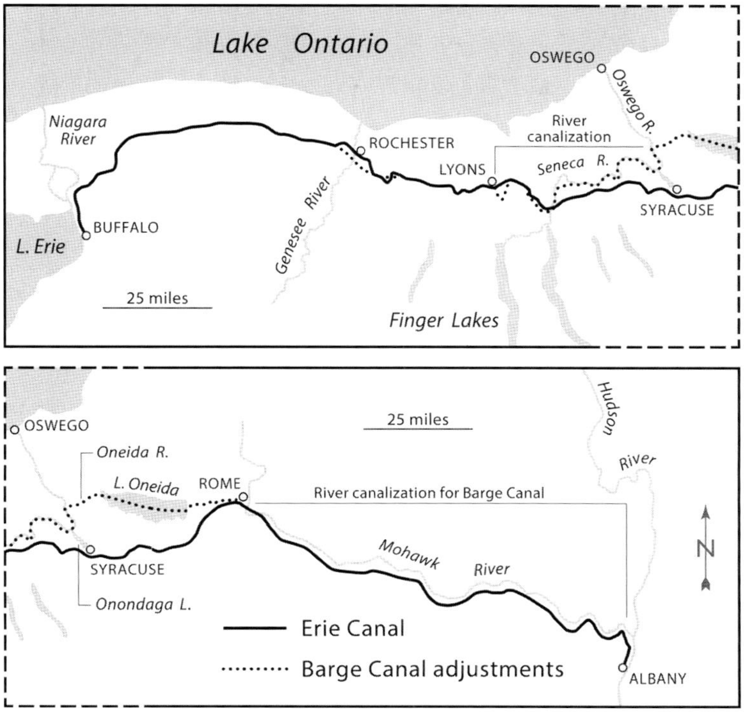

Figure 4.1. Generalized map of the Erie Canal and its successor, the New York State Barge Canal.

The Erie Canal's most vigorous and effective advocate was DeWitt Clinton (1769–1828), a New York politician who had served multiple terms as mayor of New York City and was governor from 1817 to 1822. He was one of the first appointees to the Erie Canal Commission, set up by the state legislature in 1810 to study the project's geographic and economic prospects. When Clinton pressed the legislature to fund construction with state-backed "internal improvement" bonds, his opponents dubbed the project "Clinton's Big Ditch."[6]

Construction could not proceed until field surveys compared the vaguely defined interior route with an equally unresolved path that included Lake Ontario. In 1808, surveyor-general DeWitt appointed James Geddes (1763–1838), a surveyor who lived near Onondaga Lake and what is now Syracuse, to investigate three parts of the two routes: a connection between Lakes Ontario and Oneida, a companion connection at the west end where the Niagara River descended from Lake Erie

into Lake Ontario, and an interior route that bypassed Lake Ontario altogether. All three segments were west of the broad summit near the Great Carry, generally accepted as the highest elevation on any canal between the Hudson and Lake Erie.

Geddes's reconnaissance was very much a leveling survey focused on elevation differences and lockage needs. Unnecessary intervening summits (high ground) needing additional locks for ascending and then descending were to be avoided, and wide streams and broad valleys bridged only with aqueducts or embankments were to be minimized, even at the expense of a longer, less direct route. No less important was a reliable supply of lockage water delivered from higher elevations by a network of dams, reservoirs, and feeder canals.

Though Geddes identified suitable pathways for both routes, his preference for an interior canal is apparent in warnings that the harbor at Oswego would need an expensive enlargement, that a canal connecting Lake Erie with the St. Lawrence risked losing commercial benefits certain to accrue to Albany and New York City, and that the interior route "ought to be preferred, as being free from the risk and uncertainty of wind and waves [and] that merchants can afford to pay higher freight rates when [their] property is secure, and will arrive on a day certain."[7] He also quoted an unnamed correspondent who argued "that the inland canal would always be safe in the event of a war with Great Britain"—a then-imminent event that American history texts call the War of 1812.

Although maps might suggest Oneida Lake as an obvious link between the Mohawk River and Lake Ontario (figure 4.1), a relatively direct connection between the two lakes would add an intervening summit "110 feet above the level of Oneida Lake" with a consequent ascent and descent requiring "220 feet of extra lockage."[8] By contrast, a thoroughly inland route that bypassed Oneida Lake could include "the Onondaga Salines [which] enjoy an extensive level, very valuable for the immense supply of wood it would afford." And no less beneficial to the Onondaga Lake salt industry, which emerged as an important shipper after the Erie Canal opened in 1825. Syracuse, which grew from a small village named Salina at the southeast end of the lake, became known as the Salt City.

Geddes had doubts about the interior route until a eureka moment in December 1808, when he discovered that the Genesee River (figures 4.1 and 4.2), which flowed northward toward Lake Ontario, could provide water for a canal that ran eastward all the way to the Seneca River without an intervening summit. A key obstacle was the broad valley of Gerundegut Creek (also known as Irondequoit Creek), which also flowed northward.[9] Geddes's plan called for an artificial ridge, or *embankment*, that carried the canal over the creek to a natural ridge that provided a level path across the rest of the valley into Mud Creek, which ran eastward. Known to geologists as an *esker*, this natural ridge is a long, low, winding hill consisting of sediments, principally sand and gravel, deposited by a stream flowing beneath the glacier that covered the area thousands of years ago. An esker's serpentine shape reflects a natural tendency of streams to meander, even when flowing beneath hundreds of feet of ice.

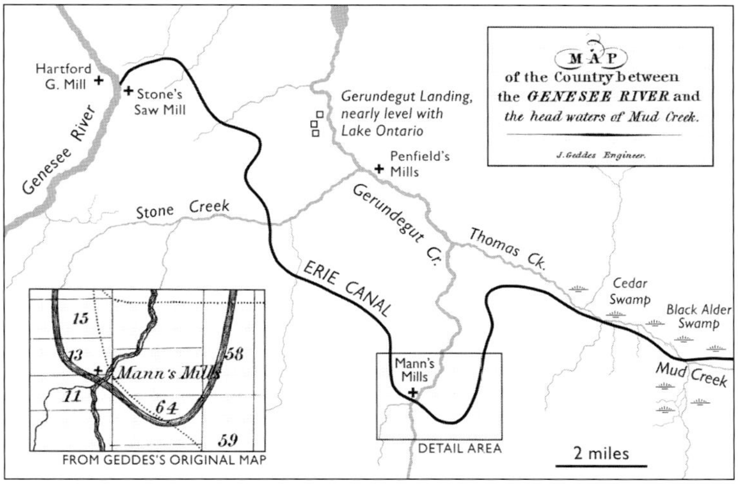

Figure 4.2. Principal features retraced from James Geddes's "Map of the Country between the Genesee River and the Head Waters of Mud Creek," which accompanied appendix C of Geddes's 1809 report to surveyor-general Simeon DeWitt. Original map, with frame, measures 8.2 × 5.4 in.

Geddes's report to DeWitt included five maps, one of which he mentioned thirteen years later in a letter to geographer William Darby.[10] I traced essential features from this map, which measures 8.2 × 5.4 in., onto the smaller version in figure 4.2, which shows the Erie Canal's somewhat roundabout route from the Genesee River, to the west, across Irondequoit (Gerundegut) Valley into Mud Creek, now called Ganargua Creek. While measuring elevation differences, Geddes discovered the possibility of a progressively downhill route from the

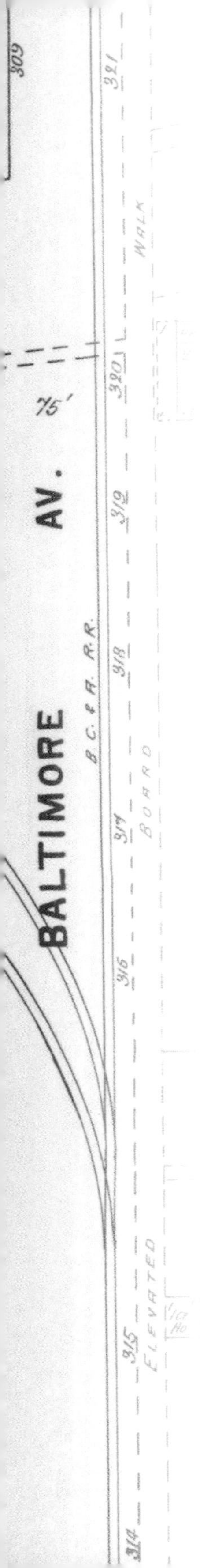

Genesee to the Seneca with no unavoidable high ground in between, and after working his way up Gerundegut Creek, he found a suitable crossing point near Mann's Mills. Multiple swings east and south from the Genesee at Stone's Saw Mill afforded a level route to Mann's Mills, where canal builders constructed a massive earthen bridge called the Gerundegut Embankment.

In his letter to Darby, Geddes described the partly completed embankment inspired by his discovery.

> The passage of the Irondequoit Valley is on a surface not surpassed, perhaps in the world, for singularity. No adequate idea can be conveyed without a map. Those ridges along the top of which the canal is carried are [in] many places of just sufficient height and width for its support, and for 75 chains [0.94 miles] the canal is held up, in part by them, and in part by artificial ridges, between 40 and 50 feet above the surface of the earth; the sides of them are in most places remarkably steep, so that when the work is finished, the appearance to the stranger will be, that nearly all those natural embankments were artificial works.[11]

His accompanying map (figure 4.3) shows a straight artificial embankment, nearly a quarter mile long, connecting gently curving stretches of the canal on opposite sides of the creek. Diverging dashed lines along the base of the embankment reflect a steady increase in height between the west end of the embankment and a stone arch across the creek. Other details include the mill building at one end of a dam across Gerundegut Creek, a large millpond for water power, relatively steep flanks where the esker rises above the floodplain, and a level avenue along the top of the esker with barely enough room for the canal and its towpath.

Geddes marveled at this achievement.

> While traversing these snowy hills in December 1808, I little thought of ever seeing the Genesee waters crossing this valley on the embankment now constructing over it. I had, to be sure, lively presentiments, that time would bring about all I was planning, that boats would one day pass along the tops of these fantastic ridges, that posterity would see and enjoy the sublime spectacle, but

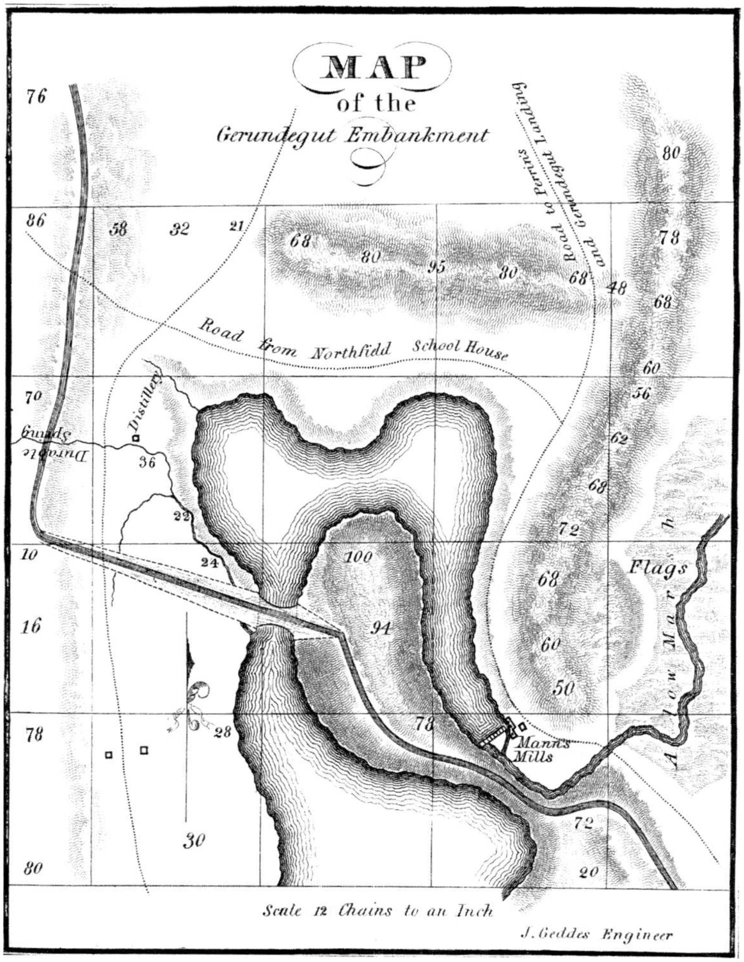

Figure 4.3. Engraving of James Geddes's map of the embankment near Mann's Mills. Its stated scale, "12 chains to an inch," refers to an image 4.6 in. wide, representing 55.2 chains; at 80 chains per statute mile, the area shown is approximately 0.69 mi. from left to right.

that for myself, I had been born many, many years too soon. There are those, sir, who can realize my feelings on such an occasion, and can forgive, if I felt disposed to exclaim Eureka, on making this discovery. .... Boats to pass over these arid plains, and along the very tops of these high ridges, seemed then like idle tales to everyone around me.[12]

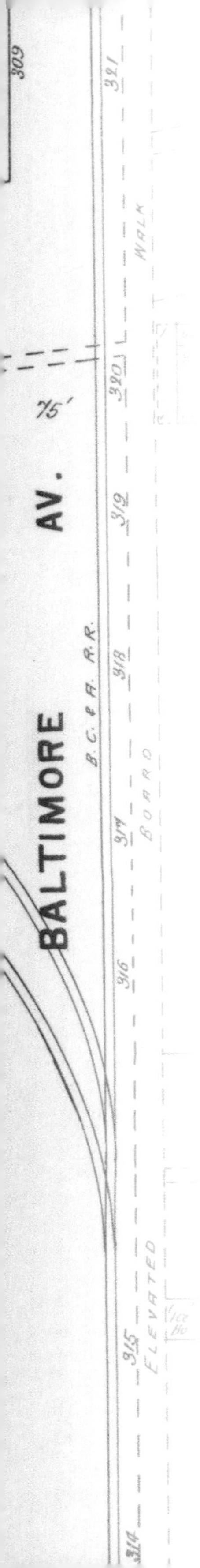

In engineering parlance, an *embankment* is a level man-made ridge, built with compacted earth and stone to carry a road, railroad, or canal across a lower area too broad for a mere culvert. Unless the crossing in question is very high or the source of fill material too distant, an embankment is usually less expensive to build than an aqueduct or a bridge, especially if it connects two cuts or excavations that provide the fill. Building the Gerundegut Embankment was particularly difficult because adequate fill was not available from an immediately adjacent cut, and suitable soil excavated miles away had to be lifted as much as 70 ft. above the valley floor.[13] Moreover, because the base of the embankment was partly on quicksand, which was known to settle, thousands of wooden piles were driven into the ground to provide a stable foundation for the fill. Although other parts of the embankment were less demanding, the heavy semicylindrical masonry culvert 245 ft. long and 26 ft. in diameter that carried the creek under the embankment required more than nine hundred piles roughly a foot in diameter and between 12 and 20 ft. long.[14]

The human effort was similarly immense, with dueling ethnic narratives. According to two recent popular histories of the canal, three thousand Irishmen using shovels, wheelbarrows, wagons, and draft animals worked on the embankment for two years.[15] Although large numbers of Irish no doubt worked on the canal, native-born and other groups probably contributed to the embankment, though how significantly is questionable. In *Rochester: A Story Historical*, published in 1884, local writer Jenny Marsh Parker (1836–1913) touted the role of area farmers but ignored the Irish.

> When the Irondequoit embankment was built, nearly every male resident of the adjoining towns turned out with pick, spade, and wheelbarrow, unless physically unable. Some of the wealthiest farmers in the county are proud of telling that they worked for seventy-five cents a day on the Erie Canal.[16]

Perhaps prone to hyperbole, Parker put the height of the embankment at 80 ft.

My attempt to portray the Erie Canal on a single map (see figure 4.1) was a frustrating and inherently flawed attempt to capture a dynamic transport network in a single graphic that describes the route, includes its principal adjustments, and accommodates the names of lakes, rivers, and cities mentioned in the chapter.[17] I compromised by dividing the canal's long and narrow horizontal route into two sections, as a graphic paragraph of sorts, with the western portion on top in accord with the customary flow of words from left to right and from top to bottom. Cramming the entire canal into a single map no longer than a line of type was impracticable—I know because I tried—and dividing the route with a single, brutally abrupt vertical line was unnecessary once I realized that modest overlap, represented by the dashed portions of the two frames, allowed the uncongested placement of competing labels near the center: a serendipitous strategy for including both the Oswego River and the city of Oswego.

Overlapping panels also enabled the labeling of tiny Onondaga Lake, otherwise edited out of the discussion. Tagging it required joining its symbol and label with a thin leader line, something I caution my map design students to avoid wherever possible. A second leader line links the short, but nonetheless notable, Oneida River with its label.

To keep the map focused on the Erie Canal, I omitted the Champlain Canal, authorized in 1817 to connect Albany with Lake Champlain, as well as the half-dozen lateral canals built with public money in the 1820s and 1830s to serve areas north or south of the main canal. The most significant of these was the Oswego Canal, which added locks adjacent to rapids in the Oswego River (see chapter 3). Although figure 4.1 ignores the canalization of the Oswego River, its opening in 1828 demonstrated the practicality of the Lake Ontario route that Geddes had rejected. A more obvious omission is the idiosyncratic, ill-conceived Black River Canal, which ran 35 mi. north from Rome through 109 locks—that's right: 109 locks, in contrast to only eighty-three locks along the Erie Canal's original 363-mile route between Albany and Buffalo.[18] Completed in 1855, largely out of service by 1900, and formally abandoned in 1925, its presence on the map would be a needless distraction.

Repeated alterations of the main route from Albany to Buffalo were a cartographic challenge. State officials addressed an unforeseen increase in traffic by widening and deepening the canal between 1835 and 1862—and lengthening its locks to accommodate longer boats. Although sections of the canal were regularly redone, to lower maintenance costs, deal with unstable soil, or replace an expedient but now dysfunctional design, the most radical change occurred between 1905 and 1918, when the state rebuilt much of the canal and renamed it the New

York State Barge Canal. Unlike the original alignment, which merely took advantage of the Mohawk River's broad floodplain, the Barge Canal replaced the river's natural channel with a less sinuous waterway of minimum width and depth. East of Rome, the canalized river was rarely more than a mile north of the original canal, but neither these disparities nor the distinction between the natural and canalized rivers merits a separate delineation.

Consequently, a single solid black line focuses on the original route while a dotted line, only slightly less prominent, points out noteworthy alterations such as the new route westward into Lake Oneida, no longer problematic because petrol-powered boats had replaced animals and towpaths. As shown in figure 4.1, the old route to the south, through Syracuse, was replaced by a new route along a canalized Seneca River nearly as far as Lyons. A connection to Onondaga Lake continued service to the Syracuse Inner Harbor, at the southeast end of the lake. Other connections allowed canal boats into several of the larger Finger Lakes. West of Lyons, the Barge Canal mostly followed the original route, except for a significant rerouting near Rochester, where the state built a bigger, better Irondequoit Embankment, sometimes called the Great Embankment.[19]

Look carefully at the upper portion of figure 4.1 directly southeast of Rochester where the solid black line makes a short southward jog, shown in greater detail in figure 4.2 as the Erie Canal's southward swing around Mann's Mills. Then note in figure 4.1 how the Barge Canal (dotted line) takes a more direct route to the northwest, in contrast to what became the "Old Canal" (solid line), which veers more northward toward (and into) Rochester. The Barge Canal brought a new aqueduct across the floodplain of the Genesee River as well as a longer, straighter, slightly higher embankment across Irondequoit (Gerundegut) Creek.

The new embankment reflects not only a legislative mandate for a wider, deeper waterway but also the emergence, during the nineteenth century, of systematic topographic maps, mechanized construction equipment such as the steam shovel, improved concrete that provided a leakproof liner for the elevated canal, and a cadre of college-trained civil engineers more professional, if not more astute, than their forebears, who mostly learned on the job and for whom the Erie Canal has been hailed as the country's first school of engineering.[20] Even so, a serious accident during the construction of the Barge Canal's new Irondequoit crossing demonstrated that engineers still had much to learn about elevated concrete troughs and groundwater.

Engineering News, a weekly newspaper for engineers of all kinds, described the incident in a seven-page story titled "Break in the New York State Barge Canal at the Irondequoit Creek Crossing."[21] On September 3, 1912, workers reinforcing the underside of the masonry culvert carrying Irondequoit Creek under the embankment noticed water streaming through a hole in the roof. Seepage from the reservoir behind the dam at the grist mill (Mann's Mills in figure 4.2) had saturated the base of the embankment, and excavation downstream from the culvert, which was to be lengthened, had increased the seepage. The culvert had been extended once before, in 1859, when the canal and embankment were widened, and the Barge Canal's wider trough required a further broadening of the embankment at its base as well as a lengthening of the culvert. Although explanations for the break vary, key factors include progressive deterioration of the culvert's wooden foundation, a gradual weakening of the masonry arch, and progressive settling of the embankment over the culvert. Collapse of the culvert precipitated the collapse of the canal's concrete trough, and water from the trough contributed to the washout of the north (downstream) side of the embankment, which carried sand a quarter mile downstream. No one died, but traffic past the break was halted for six weeks, and repairs were estimated to cost at least $200,000—equivalent to more than $5 million today.[22] Greater damage was avoided because supplemental "steel guard gates" (movable walls that could block off the trough) to the east and west of the area were lowered as soon as a break appeared imminent.[23]

In planning a reconfigured route for the Barge Canal, engineers might have taken advantage of 1:62,500 topographic maps published by the US Geological Survey, which had been mapping upstate New York in the 1890s for its 15-minute quadrangle series. Although contractors building the canal required engineering drawings at a much larger scale, cartographic treatments similar to the upper panel in figure 4.4 show the terrain around the existing canal and river. Based on an 1899 plane table survey and published in 1902, the map includes the village of Lyons, directly north of the Erie Canal. Immediately to the south is the Clyde River, which flows southeastward from the middle of the map's left edge. Feeding the Clyde from the south is the Canandaigua Outlet, which drains Canandaigua Lake.[24] Though the excerpt does not show it, the Clyde River flows southeastward

(off the map) for about two miles before turning northward across a flat, swampy area that begins at the right edge of the map—part of a meander loop is barely visible at the lower right, near the railroad track. Just off the upper-right corner of the map, the river turns eastward to feed the Seneca River, of which it is the principal tributary.

Although the Clyde River was an important source of water, the original Erie Canal took a different route. After skirting the southern edge of Lyons (which owes much of its existence to the canal), the waterway continued northeastward along a series of relatively straight stretches that wind through a group of low hills. These hills share a north–south elongation that identifies them as drumlins, formed at the bottom of a glacier that had picked up fine sediment from the floor of Lake Ontario as it plowed southward. Forced to climb progressively higher as it moved beyond the lake, the glacier dropped sediment it could no longer carry, and then overrode and molded these deposits into streamlined hills—*drumlin* is a Gaelic word meaning "small or narrow hill," and Ireland has many similar landforms. All the drumlins on the map sit on a flat, somewhat swampy plain that had once been the floor of a much larger Lake Ontario, and their shared alignment records the direction of ice flow. This lake plain provided level paths for the canal, the river, and portions of the area's three railways.

Figure 4.4's lower panel shows the same area as mapped a half century later using 1942 aerial photography and a 1953 plane table survey. Because both panels were printed in color, I used Photoshop to minimize graphic conflict before converting the scanned images to black and white. Experimentation with contrast adjustment helped emphasize canals, rivers, and railways. Because the older map's paper had yellowed, lightness and contrast adjustments avoided an overly dark background. Because the more recent map employed two additional colors—green for orchards and woodland and red for some roads and labels—color separation, filtering, and contrast adjustment removed the green tint entirely and deemphasized roads, buildings, and most labels, which remain in the background as a light-to-medium gray. On both maps, grayscale remnants of the brown contour lines are no more prominent than needed to depict the general form and presence of drumlins. A close look should reveal that terrain contours on the more recent map (bottom) are more detailed than on the earlier map (top) thanks to the greater precision of aerial photogrammetry.

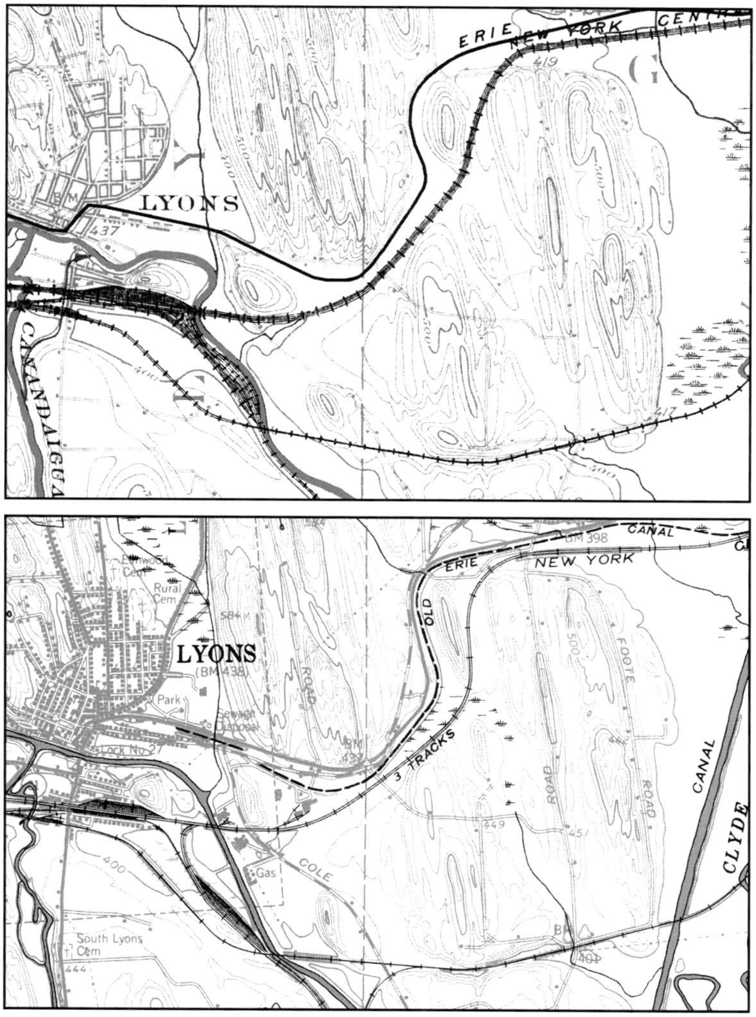

Figure 4.4. Excerpts from 1:62,500 topographic maps illustrate changes in canal and river alignment between 1899 (*top*) and 1953 (*bottom*). One inch represents one mile. The earlier map is based on an 1899 survey, and the later map reflects 1942 aerial photography and a 1953 plane table survey.

Collectively, these twin excerpts describe the impact of canalization on the Clyde River. Note on the upper map that the original canal and the river were separate waterways. By contrast, the lower map shows the river in a straighter, carefully engineered channel, confined by concrete to resist the natural tendency to meander. The lower-right corner of the excerpt reveals a divided river represented by the straight channel of the Barge Canal and parts of a few meander loops belonging to the reconfigured but relatively unconstrained Clyde River, which shares the generous floodplain. The Barge Canal's far less tortuous path reflects a dam, a lock, and a modest embankment.

Oddly, the label "CANAL" (all capitals and part of the name Erie Canal—that is, the Barge Canal) is noticeably smaller than the label for "CLYDE" (part of the name Clyde River), even though the former watercourse is wider and more economically significant than the latter. Perhaps the latter label's larger size reflects a cartographer's viewpoint that the broad floodplain belongs to a river, not a canal.

Canalization left behind an abandoned channel represented on the lower panel by a dashed line (long dashes, short gaps) with the same alignment as the canal in the upper panel but now labeled "OLD ERIE CANAL" (all capitals in the same font used for the Barge Canal). Where not filled in or otherwise appropriated for a road improvement or urban development—as within the village of Lyons—the ditch remains as a historical artifact, sometimes with its towpath preserved as a hiking trail or bike path. The Old Erie Canal in figure 4.4 is part of a recreational network known as the Erie Canalway National Heritage Corridor, a public-private partnership affiliated with the trail advocacy organization Parks and Trails New York as well as with the National Park Service and the New York State Canal Corporation, which operates the Barge Canal. Wayne County, which includes the area in figure 4.4, also has a stake in the endeavor, which promotes recreation, tourism, and regional history.

Several trail and tourist maps show the abandoned canal east of Lyons. On a large folded map and brochure that beckons viewers to "Experience New York's Legendary Canals," the roughly four-mile stretch of Old Erie Canal is represented by a dashed line that diverges away from the Barge Canal east of Lyons, but this symbol's northward bulge is mostly obscured by large numbered dots that call attention to two local attractions: Historical Murals of Lyons (a cluster of a dozen outdoor murals) and Trail of Hope (a quarter-mile nature trail). By contrast, the old canal's location and configuration are clear in the Wayne County Parks and Recreationways Master Plan, in which a pair of detailed maps point out two nearby parks along the route and an "existing multiuse trail."[25]

Anyone seeking details and directions is better served by the interactive *Erie Canalway Trail Map* maintained by Parks and Trails New York. Hosted on a Google Maps™ platform, the map can zoom out to encompass the corridor's 524 mi. of navigable waterways and adjacent trails, or zoom in to show street names, parking areas and trail heads, tourist attractions, and bike shops.[26] Figure 4.5, redrawn for clarity in black and white, illustrates how the map describes the trail surface as *stone dust* (crushed limestone), *natural* (old towpath), or *on-road* (a designated or implied bike lane on an existing public road paved with asphalt). Instead of a bar scale, the map has a measuring tool so users can estimate cumulative distance along a chain of straight-line segments. In addition to showing connections to other trails, the map lets users search for restaurants, grocery stores, lodging, ATMs, and other amenities.

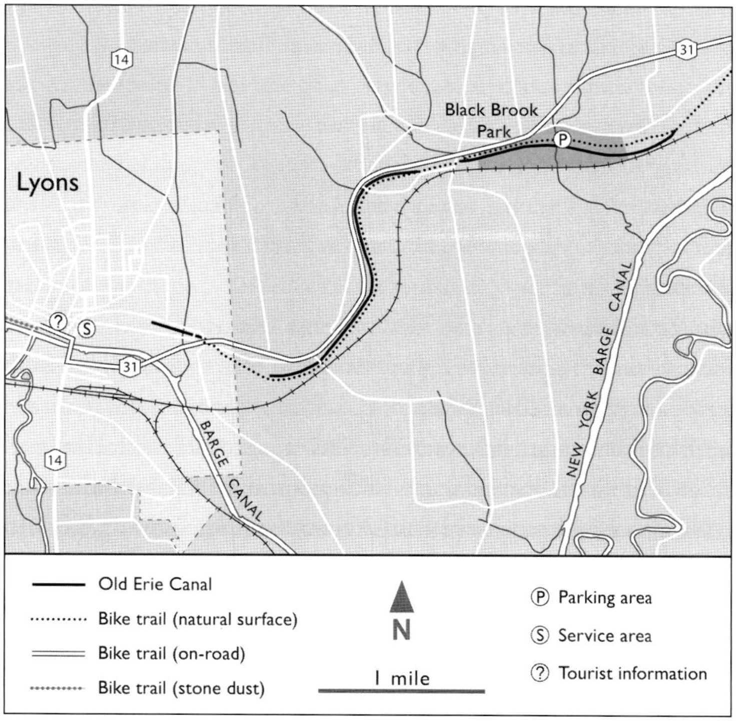

Figure 4.5. Snapshot of Erie Canalway's interactive trail map for the area east of Lyons, New York.

Recreation and tourism now account for the continued existence of the Barge Canal, a massive upgrade built a hundred years earlier to compete with the railroads. Aside from unusual shipments such as windmill blades too cumbersome for rail or highway transport, the canal mostly serves recreational and tourist users, who bring canoes and kayaks or take a one-hour or half-day cruise.[27] You can launch your own boat or rent one, enjoy a dinner cruise, or transit the canal on a two-week luxury Great Lakes cruise from New York to Chicago. In 2017, the State Canal Corporation, which maintains the waterway and staffs the locks, waived the "all tolls or fees for recreational vessels .... in recognition of the 200th anniversary of the start of Erie Canal construction"—not a great loss in revenue because an annual pass for a private cabin cruiser normally costs only $100.[28] Also free are the numerous hiking and bike trails that line sections of both the old canal and the Barge Canal.

Parks and Trails New York is committed to closing remaining gaps in the 360-mi. Erie Canalway Trail between Albany and Buffalo, which was 80 percent complete in 2015.[29] Recently funded segments include the one-mile gap between the stone dust and natural-surface bike trails in the village of Lyons (figure 4.5). Periodically updated maps describe incremental progress in completing the old canal's recreational network. Because extensive transport networks of all types typically evolve step by step, maps have always had a role in advocacy, news reporting, and historical narratives.

A good example is the New York Central Railroad, which roughly parallels the Erie Canal in figure 4.4. The prominent east–west rail line (four tracks in 1899 and three in 1953) reflects the 1853 consolidation of railroads linking Albany with Schenectady, Schenectady with Utica, Utica with Syracuse, Syracuse with Rochester, and Rochester with Buffalo. Now part of CSX Transportation, its two tracks handle Amtrak® service as well as numerous long freight trains. A lesser east–west line, farther south in the excerpts, was built in the 1860s and 1870s to compete with the New York Central. Named the West Shore Railroad because it followed the west side of the Hudson River north from New York to Albany, it was absorbed into the New York Central in 1885 and operated as a second-string carrier until the 1970s, when large sections were abandoned or sold as independent so-called "short lines." Its omission from figure 4.5 reflects the abandonment in 1959 of a relatively level roadbed that would have been a prime candidate for a rail trail were not the old canal's towpath more scenic.[30] The third rail line in figure 4.4 runs from Lyons south to Geneva and Corning and is now a branch line on the Norfolk Southern system.

Aside from a shared appreciation for level pathways, like those in figure 4.4, canal and railway engineers differed significantly in their appreciation of rivers and streams, which were an essential source of water for canals but generally a nuisance for railways because of the expense of constructing bridges, filling in well-drained paths across swamps, and protecting tracks from flooding and erosion. Although canals required aqueducts and culverts where a stream crossing was unavoidable, railway planners had to assess construction and operating costs for more numerous alternatives, some of which might require cuts through low hills or tunnels through ridges. For example, the West Shore Railroad's relatively straight stretches southeast of Lyons required several cuts and fills, identified in the lower panel of figure 4.4 by their *hachures*, a cartographer's term for short ticks, perpendicular to and evenly spaced along contour lines and pointing downward. The surveyors who laid out the older railroad avoided these complications by mimicking the old canal's route across the drumlin field. By contrast, the West Shore's surveyors, by more aggressively reshaping the terrain, devised an alternative route that was only slightly longer. A half century later, much improved earth-moving technology fostered a notably wider and straighter Barge Canal.

Reconfiguring the river to accommodate the Barge Canal required engineers to become astute cartographers and geographers, ever aware of differences in soil permeability and elevation as well as in the volume of earth to be scraped out, transported, and deposited in a raised bed or embankment. Locating and designing a railroad demanded similar attention to the quality and quantity of relocated surface material, and when cuts did not yield sufficient fill, railroad builders relied on ad hoc excavations euphemistically called *borrow pits*, as if the loan would someday be repaid. Where cheap timber was available, long trestles afforded a less expensive strategy for crossing broad valleys.

Settlement of the American West, widespread demand for rail service, and optimistic investors helped fuel an expansion of the rail system from 9,021 to 193,346 route miles between 1850 and 1900, while the total length of American canals dropped from 5,460 to 2,515 route miles between 1860 and 1880.[31] Consolidation in the rail industry led to standard time zones as well as a standard

gauge (distance between the rails) of 4 ft., 8½ in., which promoted the efficient interchange of freight cars between connecting companies. Even small rail yards such as those in figure 4.4 made junctions like Lyons important as nodes in an integrated national network.

Rail expansion depended on engineering experts and educators like William Galt Raymond (1859–1926), whose career mixed collegiate training with practical experience in the railroad industry.[32] His résumé includes summer and intermittent full-time employment while working on a bachelor's degree in civil engineering, awarded by Washington University in 1884, and various consulting gigs while teaching at the University of California (1884–90), Rensselaer Polytechnic Institute (1892–1904), and the State University of Iowa (1904–26), where he was appointed dean of the College of Applied Science in 1905. In 1890, Raymond coordinated location surveys for a proposed line through California's Coast Range that would connect the Atchison, Topeka & Santa Fe with San Francisco. Although the project was shelved when financing failed, it informed his classic textbook *Elements of Railroad Engineering*, first published in 1908, and kept in print through five subsequent editions, the last two published posthumously, in 1937 and 1947.[33]

As its title implies, Raymond's textbook covered a broad range of issues faced by the industry, from rails and ties to traffic generation and government regulation, from sidings and signals to locomotives and trackside buildings, and from operating expenses to planning. Roughly a quarter of its more than four hundred pages addressed surveys and maps, with separate chapters for reconnaissance, preliminary surveys, location surveys, construction surveys, and betterment surveys. The last category encompassed strategies for tweaking parts of an existing rail line to reduce costs and increase speed and capacity. Tweaking could be costly but necessary if the original line had curves too tight for the newer, larger locomotives used for high-speed trains.

Key considerations in planning a section of track are its curvature and grade, concerned respectively with horizontal and vertical change in alignment.[34] Curvature was treated as the radius of a circular arc to which a straight section of track, or *tangent*, was joined: the larger the radius, the smoother the transition and the less jerky the ride. Grade was measured as either the angle of slope or the change in elevation along 100 ft. of track. Potentially problematic was the line's "rise and fall," in particular the "undulation" arising when a grade that was not steadily upward or downward interfered with the uniform work of a steam locomotive, thereby increasing the operating cost.[35] The locations of bridges or curves requiring reduced speed were important, and because trains had to stop and start up again at stations, engineers preferred to place depots at summits or other high points along the line. No less significant was the *ruling grade*, defined as the steepest grade on a line. Because steep uphill grades often required additional, or "pusher," engines, concentrating these steeper grades on a single part of the line might avoid the added expense of deploying pushers (and their crews and servicing facilities) to multiple locations.

"Reconnaissance," the longest of *Elements*' five chapters, addressed the broad-brush geography of probable traffic and terrain—understanding what points the route had to connect and discerning what obstacles intervened. Raymond called reconnaissance "the most important part of railroad location [and] perhaps the most important item of railroad engineering."[36] Getting it right the first time was crucial because "a mistake in selecting the route, ... if not discovered and rectified before construction, is likely not to be rectified for many years."

In assessing the terrain's opportunities and constraints, engineers were encouraged to take advantage of existing maps, especially USGS topographic maps, with elevation contours useful for plotting vertical profiles and interpreting the configuration of landforms and drainage.

> The contour map shows readily enough what the possibilities may be. One or more routes joining the proposed termini may be selected as the most feasible of the routes of reasonable directness. The probable ruling grade may be approximately determined in ordinary country from such maps, the distance and probable amount of curvature may be known with some degree of accuracy, but the character of the material is not shown, nor the minor irregularities. A horseback trip, or possibly a trip with a team, with short excursions from the road on foot or horseback at various points, will sufficiently determine the characters of the routes so that a choice may be made from the map.[37]

Although USGS topographic maps were inexpensive (five cents a sheet) and well suited for planning scouting trips, "only a limited portion of the country has been mapped in this way."[38] The "next best" sources were county maps, typically published as wall maps covering an entire county or atlases that divided a county into rectangular sections, each on a separate page. If county maps were unavailable, plat maps, describing property boundaries as well as roads and streams, could be useful.

Raymond illustrated how simple maps of the stream network could suggest promising sites for a field excursion. The lines in the left part of figure 4.6 are streams, the arrows indicate direction of drainage out of the area, and points *A*, *B*, and *C* are near divides between watersheds draining either east or west. Where an eastward draining stream is parallel and close to a westward draining stream, as at *A*, a plausible interpretation is that the streams are separated by a "high narrow ridge [punctuated] with low saddles or passes," one of which might provide a workable east–west route through the area.[39] A narrow ridge was valuable because it might allow a short tunnel. The map also suggests looking near *B* for other passes, "narrower and easier of approach," whereas *C* is most likely a high, peaklike area with streams radiating in multiple directions, and *D* marks a comparatively level area, with hills on all sides. By contrast, the configuration of eastward- and westward-flowing streams in the right part of figure 4.6 suggests a sinuous ridge marked by the dashed line. A short tunnel at *C* or *D* might provide an efficient east–west link between lines following adjoining streams east and west of the ridge line.

Figure 4.6. William Galt Raymond showed how drainage patterns on maps without contours could suggest promising routes across a region.

Engineering students could find a dramatic illustration of these principles in a textbook published in 1892 by Thomas Cleemann (1843–93), a civil engineer in Philadelphia who had worked for the Pennsylvania Railroad and consulted on railroad projects in Peru and Ecuador.[40] Drawn in 1881 with hachures rather than contours, the map in figure 4.7 describes the Western North Carolina Railroad's circuitous crossing of the Blue Ridge Mountains. Though disappointed by the lack of contours, Cleemann marveled at "how much advantage has been taken of lateral valleys to increase the length of line." Distributing the rise in elevation over a longer distance reduced the grade, as did the route's strategically placed tunnels.

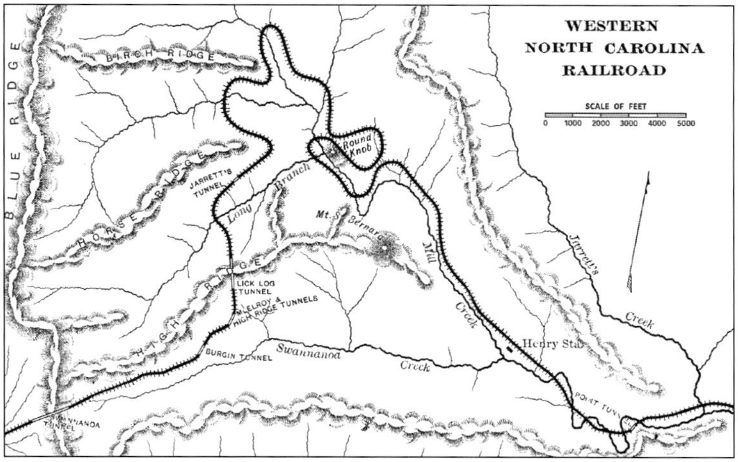

Figure 4.7. A circuitous route through the Blue Ridge Mountains kept the grade manageable by taking advantage of linear valleys. From Thomas M. Cleemann, *The Railroad Engineer's Practice* (1892).

Raymond's text included several representative examples contributed by engineering acquaintances such as W. D. Taylor, chief engineer of the Chicago & Alton Railway, who provided a 7 × 50 in. foldout describing two proposed branches, 19.6 and 9.5 mi. in length. The excerpts in figure 4.8 describe one of the two more interesting parts of the longer branch, which crosses terrain much flatter than the Western North Carolina route. Reproduced here at the same size as in Raymond's book, the profile (top) approximately matches the map (bottom), which mostly shows streams crossed along the proposed route, represented by the confined dashed line. The map also includes section lines, a mile apart, from the US Public Land Survey grid. The profile is framed by a grid with vertical lines every 2,700 ft. (slightly less than a half mile) and horizontal lines every 15 ft., which significantly exaggerates the slope of the land, represented by the heavy line just above the parallel-line shading. By contrast, the track profile, represented on the same graph by relatively long straight sections that join where there is a change of grade, indicate a cut where the track is below the terrain profile or by a bridge, trestle, or fill where the terrain is lower. Vertical labels above the profiles identify crossings by name (e.g., the West Fork of Shoal Creek) or type ("run," usually a small unnamed stream).

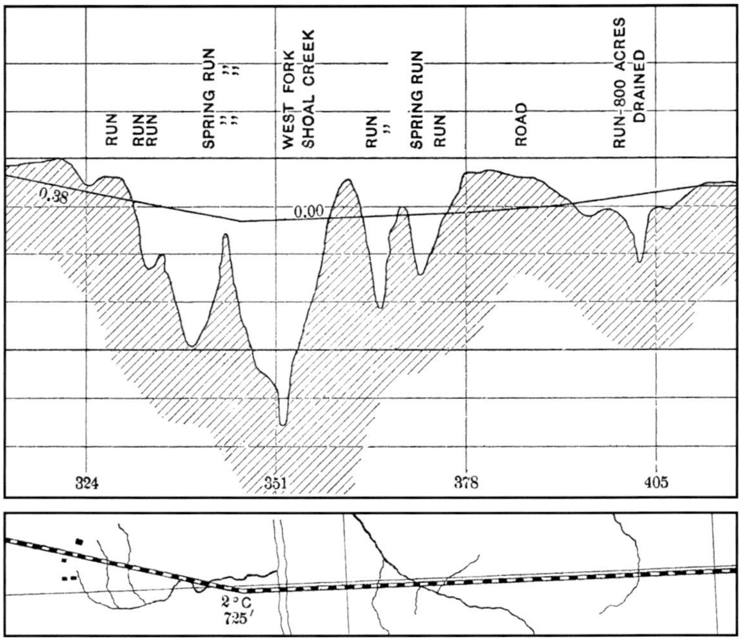

Figure 4.8. Portions of "A Condensed Map and Profile to Accompany a Chief Engineer's Report," included as a foldout in Raymond's textbook.

Careful inspection of the two excerpts reveals small numbers describing key characteristics of the route. The numbers 0.00 and 0.38 along the track profile (top) are estimates of grade resistance, which reflects the work of the locomotive.[41] In the lower excerpt, 2°C 725' indicates a 725-ft. section with two degrees of curvature between tangents. A one-degree curve refers to a circle in which a chord of 100 ft. subtends a central angle of 1°—that is, a circle with a radius of 5,729.65 ft. Civil engineers use 100 ft. as the standard for roads as well as railways. Because the radius R of a circle with D degrees of curvature can be calculated as $5{,}730/D$, a ten-degree curve, which Raymond called "relatively sharp," would have a radius of 573 ft.[42] If a one-degree curve can be described as "fairly flat," a two-degree curve, with a radius of 2,865 ft., would still be rather gentle.

For an example of ten-degree curves, engineering students could turn to William Beahan's *Field Practice of Railway Location*, which Raymond recommended as a supplement to his geography chapters.[43] The best cartographic illustration I have seen is figure 4.9, which I scanned and cropped from a plate in Beahan's book and enhanced in Photoshop to focus on the Southern Pacific Railway's Tehachapi Loop in the Tehachapi Mountains in south-central California, a favorite of model train enthusiasts and railway photographers, who can capture the front and back of a long freight train in a single shot.[44] Built out of solid granite with picks, shovels, and blasting powder, by more than two thousand Chinese laborers working for two years, the loop reduces the grade over Tehachapi Pass by substantially lengthening the line. Note that the two markers for 11 and 13 mi. are just a little over a half mile apart, and that the markers for 12 and 13 mi. are little more than an eighth mile apart. Located in 1875, the loop was the first of its kind in North America.[45] In 1998, the American Society of Civil Engineers named it a Historic Civil Engineering Landmark.[46]

Railroads continued to expand through the beginning of World War I, when the national network of freight and scheduled passenger service, however infrequent, reached nearly every town and city in the United States, and a secondary network of streetcars, rural trolleys, and high-speed electric interurban railcars extended passenger service to the growing suburbs and nearby villages.[47] But by the 1920s, an ominous contraction had become apparent. Mass production was not only making motor cars affordable to millions of working-class, middle-class, and farm families but also helping trucking companies capture much of the railroads' less-than-carload freight business. Secondary effects emerged as rising automobile ownership fueled the demand for road and street maps, and the Good Roads Movement, initiated by bicyclists in the late nineteenth century, emerged as a formidable lobby for improved highways.[48]

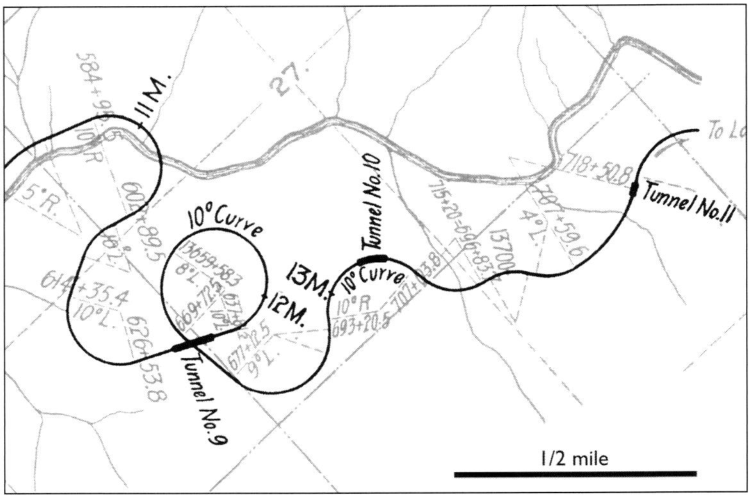

Figure 4.9. Tehachapi Loop, on the Southern Pacific Railway north of Mojave, California, between Los Angeles and San Francisco. From William Beahan, *Field Practice of Railway Location* (1904)

Responding to public demand, state governments established highway departments to locate and design intercounty roads, award contracts, and supervise construction and maintenance. Congress had recognized a federal role in highway development by establishing a series of ever larger agencies, starting in 1893 with the modestly funded Office of Road Inquiry, in the Department of Agriculture, and culminating in 1966 with the creation of the Federal Highway Administration, which uses the federal gasoline tax to provide financial and technical support for state and local projects.[49] Key legislative milestones are the Federal Road Act of 1916, which established 50/50 cost sharing for the Federal-Aid Highway System, and the Federal-Aid Highway Act of 1956, which launched the Interstate Highway System.[50]

Civil engineers were involved from the start. With long experience in designing bridges and laying out rail lines, engineers took the lead in setting standards for paving materials, traffic control, and road location. In 1919, the first edition of the *American Highway Engineers' Handbook* confirmed that the locational requirements of highways are "similar at least to those affecting the location of railways."[51] Although minimizing distance and the number of stream crossings while avoiding excessive grades and sharp curves were shared concerns, "the limitations imposed by engines and cars running on rails are much narrower and more severe than in the cases of the more flexible vehicles using roads and streets." Indeed, cars and trucks never required helper vehicles and were immune to the adverse effects of undulation on a train of railcars. Another key difference was the need to quickly drain water from impervious pavement, hardly a problem for a rail line's ballasted roadbed. Moreover, highways not only required a minimum gradient, to prevent ponding of storm runoff, but needed to expeditiously divert roadside drainage ditches into natural streams and provide city streets with their own sewer system. Hardly surprising that a speaker addressing a 1914 Good Roads symposium on "Road Surveying and Mapping" noted that "a map of the territory, especially a United States Topographic map, will always be helpful."[52]

By the 1950s, highway engineers were well aware of the advantages of aerial photography, which had rapidly revolutionized topographic mapping.[53] Used during both World Wars for identifying targets and directing troop movements, air photos had acquired a broader utility with the development of photogrammetric plotting instruments called *stereoplotters*, which exploit the principle of stereoscopic vision, whereby the human eye-brain system perceives a three-dimensional image when viewing pictures of the same scene photographed from different positions. A pair of overlapping air photos taken from a similar altitude provides a three-dimensional model of the terrain that allows a cartographer or engineer to measure elevation differences or draw elevation contours more accurately than with a plane table. In the decade following World War II, the Geological Survey's topographic division completed a tumultuous transition from field-based mapping to photogrammetry, and highway engineers realized that air photos were not only a valuable supplement to contour maps but also a potentially richer source of information.[54] In 1960, William Pryor, chief of the Aerial Surveys Branch in the Bureau of Public Roads, summarized the value of overhead imagery for locating highways in a four-step diagram (figure 4.10) that used three terms prominent in railway engineering: *reconnaissance*, *preliminary survey*, and *location survey*.[55]

Integrating air photos with each other, as well as with existing maps, relied on another cartographic innovation: plane coordinate systems, which assigned all point locations (such as road intersections and the centers of individual photographs) a pair of rectangular coordinates, known as *X* and *Y* in college geometry and as *eastings* and *northings* in cartography. Rectangular coordinates were

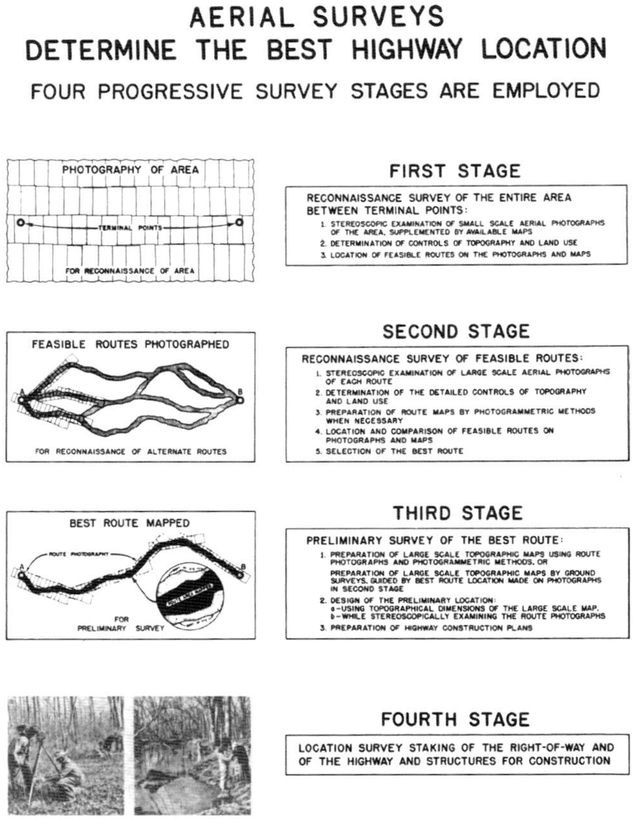

Figure 4.10. William Pryor's four stages of highway location reflect the versatility of aerial imagery.

especially useful for calculating distances when the user could tolerate minor inaccuracies associated with earth curvature. These minor errors were acceptable to the military because expeditious calculations allowed artillery officers to reliably target positions apparent on a map but too far away to see.[56] Error was unavoidable because the coordinates were tied to a map projection, on which distortion was unavoidable. The most suitable projections had a mathematical property known as *conformality*, which meant (in theory) the absence of local distortion of angles and directions—a useful property when aiming a long-range cannon. If the area covered was not vast, the error resulting from projecting the curved earth onto a flat sheet of paper was typically less than the error involved in estimating a location's coordinates from grid lines printed on the map.

Of the various grid systems available, engineers preferred the state plane coordinate system (SPCS), implemented in the 1930s at the request of a North Carolina highway engineer.[57] The requirement that distances calculated using SPCS coordinates differ from distances on a curved earth by no more than one part in 10,000 requires the division of some states into multiple zones.[58] New York, for instance, has four zones (figure 4.11). On USGS topographic maps, the SPCS grid is represented by tick marks in the map's margin, ostensibly to avoid clutter. As described in figure 4.11, lines 10,000 ft. apart connecting tick marks on opposite sides of the map are part of a rectangular grid integrating all map sheets in the zone into a single computational network. A map sheet's SPCS grid lines typically look tilted, as shown, because the rectangular grid superimposed on the projection for the zone aligns perfectly with the earth's meridians and parallels only at the center of the zone.

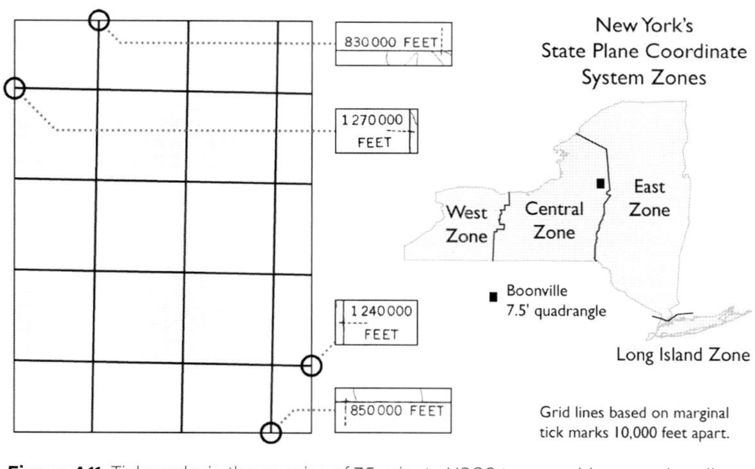

Figure 4.11. Tick marks in the margins of 7.5-minute USGS topographic maps describe a 10,000-ft. grid based on the state plane coordinate system, shown here for the Boonville, New York, quadrangle, in the state's central zone. Facsimile images show the single numbered tick mark on each edge.

Neither SPCS grid lines nor air photos were available in the late nineteenth century when the emerging electrical grid posed a new challenge for engineers. Early twentieth-century textbooks and handbooks covering electrical transmission lines reveal little interest in USGS topographic maps, except to suggest plausible locations for hydroelectric generating stations.[59] Even though Alfred Still, an electrical engineering professor at Purdue University in West Lafayette, Indiana, recognized the value of maps for avoiding "natural obstacles" or suggesting that "a slightly circuitous route may take the transmission line near to towns or districts where a demand for power may be expected in the near future," he said nothing about USGS products in his 1913 textbook.[60] By contrast, Ruben Lundquist, a consulting engineer in Minneapolis, was an enthusiastic supporter. Introducing the second paragraph of his 1912 text *Transmission Line Construction: Methods and Cost* by asserting "the first step preparatory to the construction of any kind of a line is the making of a general map," Lundquist argued that USGS topographic maps "should be secured if they are available" because the details provided "will help greatly in studying the topography of any section of the country."[61] Although the elevations might not be as precise as engineers require, "the rough and hilly sections will be easily located and other general knowledge of the topography of the country secured." But because "only a small percentage of the area of the states has been mapped in this way," plat books might be a necessary substitute.

Whether USGS maps informed the routing of early electrical transmission lines in upstate New York is unclear, even though the transmission lines in figure 4.12 mostly traverse areas for which a topographic map was published before 1907. The excerpt (adjusted in Photoshop to highlight power lines) is from a color map published between 1906 and 1912 by the Matthews-Northrup Works, a prominent map engraving firm in Buffalo.[62] Although the map lacks a publication date, online catalogs of various libraries with a similar item report publication between 1906 and 1912—Matthews-Northrup might have updated its plates over this period. I suspect the latter date is reliable for Syracuse University's copy, which I scanned, because the Salmon River Reservoir, near the upper-right corner of the excerpt, was finished around 1913, after the power plant was completed.[63] And the four 15-minute quadrangles that include the reservoir (not shown) and areas to the west, south, and southwest were published in 1905, 1895, 1906, and 1905, respectively.[64] Moreover, the quadrangle for Syracuse, where many of the lines converge, was published in 1898.[65]

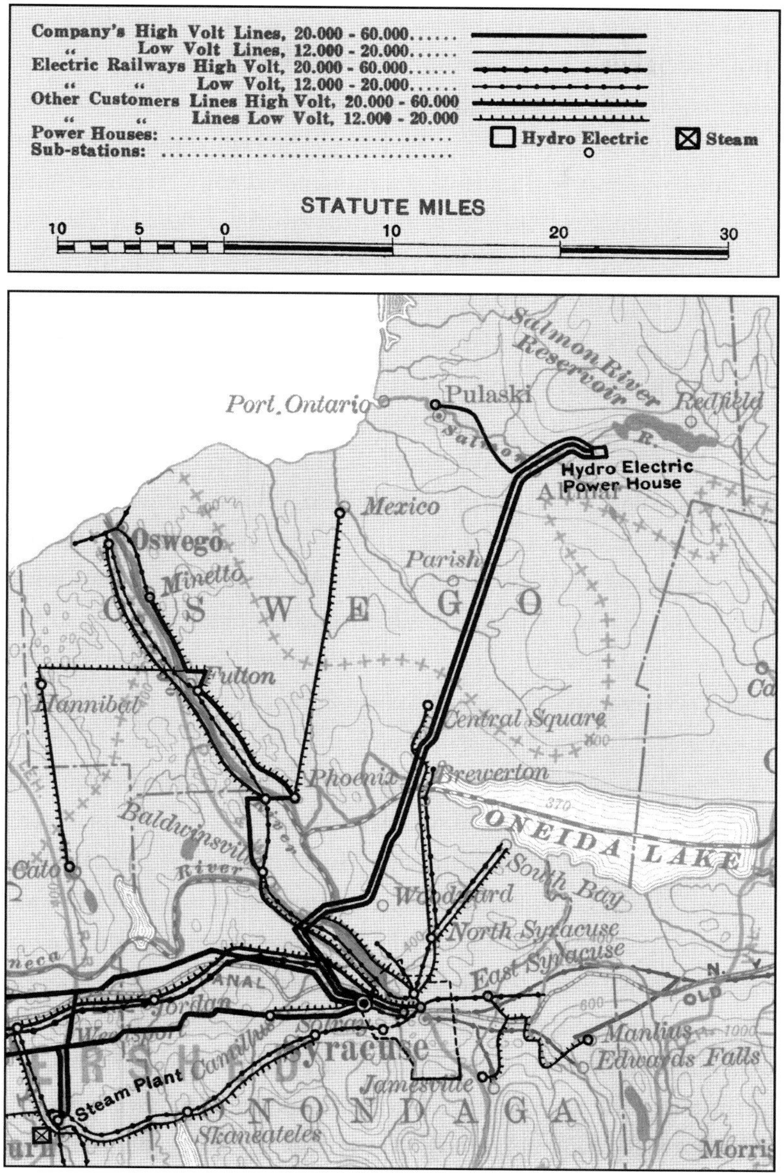

Figure 4.12. Excerpts of part of the key and the area north of Syracuse, New York, from a map (ca. 1912) of electric power lines operated by the Niagara, Lockport, and Ontario Power Company and its subsidiaries.

Whatever source materials were used in locating the high-voltage lines represented by the thicker black lines in figure 4.12, the routes were appropriately direct, to control construction costs as well as minimize the energy lost due to the voltage drop related to distance. As Worcester Polytechnic Institute engineering professor Carl Meyer observed four decades later, contour maps were especially useful in positioning a line's towers so that intervening terrain would not conflict with the minimum vertical clearance between the ground and the lowest conducting cable strung between consecutive towers (figure 4.13). Engineers had to consider not only the line of sight between towers but also the "maximum cable sag," related to the length and weight of the conductor (cable), temperature, wind, and ice accumulation.[66]

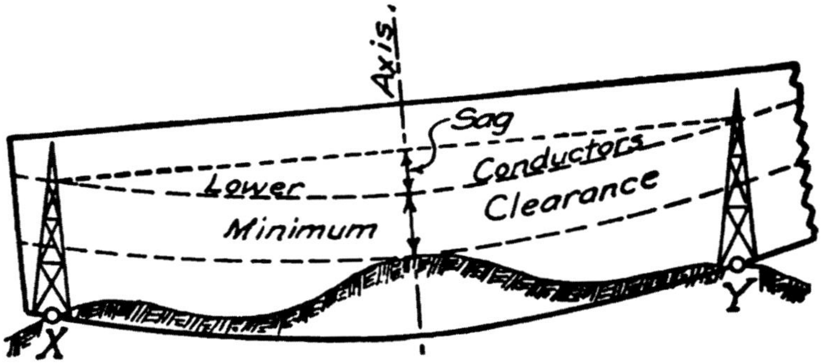

Figure 4.13. Carl Frederick Meyer illustrated how power line sag between towers made it necessary to screen potential routes for intervening terrain.

Meyer, who addressed highways, railways, and other networks in his textbook *Route Surveying*, believed strongly in the value of contour maps in planning any route sensitive to grade. He considered "an accurate contour map .... indispensable" for helping an engineer "study various locations in a small fraction of the time required for a field party to survey the lines on the ground" as well as "supply[ing] visible evidence that no better line has been overlooked."[67] He added an important caveat:

> No contour map, no matter how accurate it may be, can impress upon the mind more forcibly than field examination such details as the true significance of length and depth of cuts and fills; the nature of the materials and foundations; susceptibility to slides, snow drifting, and other maintenance difficulties; or the aesthetic values of the projected location.[68]

Meyer covered transmission lines in his final chapter, along with telecommunication lines, pipelines for oil and other liquids, natural gas pipelines, and underground conduits.[69] Although topographic maps and air photos were no doubt useful in locating these facilities, engineering students were apparently better served by a focus on established principles for routing highways and railways, the dominant infrastructure in the late 1940s, and engineers who worked in these presumably tangential industries did not need to be told, for example, that aerial surveys could expedite the routing of interstate pipelines or that fluids forced through a small-diameter steel pipe were less sensitive than trucks and trains to grade and undulation. In almost all cases, existing maps and imagery could be useful in screening pathways, while field surveys showed contractors where to dig and fill.

By century's end, electronic maps had taken on roles that Meyer had not foreseen. Digital elevation maps developed as part of the national cartographic infrastructure had become useful in locating telecommunication towers and other parts of the cellular telephone network (which threatened to undermine its landline predecessor). Cost-conscious utilities not only converted blueprints and paper maps to electronic formats, so that they could more reliably maintain hidden lines, but also joined "one-call" centers, which anyone planning to excavate was required to call. A computer converted the caller's address to an SPCS location, and if a vulnerable network was nearby, dispatched a field crew with a metal detector and a set of little color-coded flags: red for electric power; orange for communications; yellow for natural gas, oil, or steam; green for sewers and drain lines; and blue for drinking water.[70] In showing contractors where not to dig, cartography had acquired a key role in protecting buried infrastructure. The following chapter addresses another kind of protection: the weather service networks that warn of severe storms.

5

TELECOMMUNICATIONS

Telecommunication networks that rely on postal delivery, the electric telegraph, or wireless telephony are adept at both making and disseminating maps. Carrier routes, buried cables, or overhead wires linking widely separated points provide data for maps that can be distributed via another (or perhaps the same) network. These telecommunication networks might have been created originally for other purposes, and their use in making maps might depend on an institutional network created to repurpose the original messaging system. A prime example, and the focus of this chapter, is the weather map and its diverse forms, which are essential to understanding the atmosphere and forecasting severe storms.

Few maps are as diverse and ubiquitous as the synchronous weather map. Synchronous is important here because the state of the atmosphere changes by the hour, if not by the minute, and weather maps must be snapshots, based on measurements taken everywhere across a broad region at the same time. One of many modern-day examples is the cloud-cover satellite image, captured by a complex digital camera held aloft at an altitude of 22,300 mi. above a constant position on the equator, in a geosynchronous orbit, which guarantees it will not fall to earth or fly off into space. The cloud-cover image is a wide-area snapshot of atmospheric moisture, transported by global winds and often precipitated as rain or snow. Its tonal contrasts and patterns can reveal large rotating storms, and a succession of images can track the emergence and progress of hurricanes so that coastal communities are not caught off guard, as happened in September 1900, when the storm made famous by Erik Larson in *Isaac's Storm* killed over six thousand people in Galveston, Texas.[1]

Curiously, the first synchronous weather map ever drawn was based on data collected more than three decades earlier, in the late eighteenth century, by a network of observers scattered across northern and central Europe, with outliers in Massachusetts, Greenland, and the Ural Mountains in Russia. The map was drawn around 1816 by Heinrich Wilhelm Brandes (1777–1834), a physics professor at the former University of Breslau in Wrocław, Poland, who believed a map might show meaningful relationships among geographic patterns of temperature, pressure, precipitation, and sky conditions.[2] Brandes's map was never published, nor has a copy been recovered from an archive, but in a short article published in 1817 in *Annalen der Physik*, he described a plan to make daily maps of Europe for an entire year.[3] And in a follow-up article published in 1819, he described revealing patterns of barometric pressure on March 6 and 12 of 1783, when severe storms raged over parts of Europe.[4] Because there's little question that Brandes actually mapped the data, subsequent authors commemorated his achievement by using identical measurements to reconstruct his pioneering depiction. Figure 5.1 shows a reconstruction published in a 1905 meteorology textbook.[5] Wind arrows portray counterclockwise circulation around a center of low pressure, marked here by lines portraying equal departures from normal pressure—the earliest approximation of the isobars on today's weather maps.[6]

Brandes constructed his maps from data published by the Palatine Meteorological Society, founded and funded by Karl Theodore (1724–99), "elector" (head of state) of Palatinate-Bavaria, in southern Germany.[7] The society recruited knowledgeable and committed observers, supplied them with calibrated thermometers and barometers as well as instructions for constructing rain gauges and wind vanes, and encouraged them to take measurements three times a day (at 7:00 a.m., 2:00 p.m., and 9:00 p.m.) and record their observations on printed forms mailed from society headquarters in Mannheim, Germany. Observers returned the forms periodically, along with information on local births, deaths, disease, bird migration, and the growth of crops. The network encompassed thirty-seven observers but some participated for only a year or two.[8]

Brandes must have been delighted to discover the society's data in annual volumes titled *Ephemerides* (Latin for "newspapers") covering the period 1781 through 1792. Meteorological historian John Kington, who reconstructed daily weather maps for his 1988 book *The Weather of the 1780s over Europe*, relied on

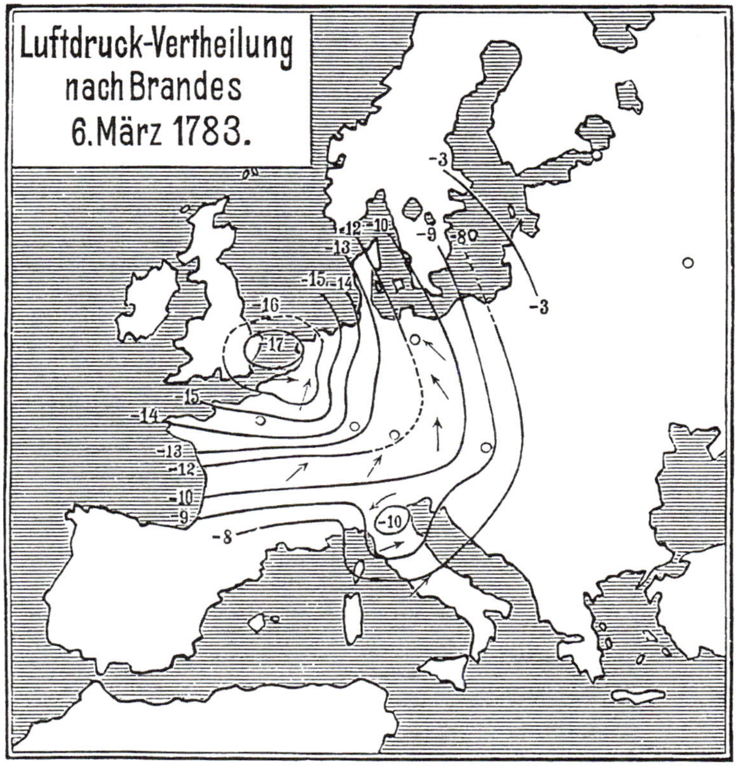

Figure 5.1. Brandes's weather map for March 6, 1783, as reconstructed for a 1905 textbook published in Germany.

the *Ephemerides* as well as other observers, including the slightly denser network of the Société Royale de Médecine, the records of which had been buried in a Paris archive for nearly two centuries, following the death by guillotine of the society's founder, Antoine-Laurent de Lavoisier, in 1794.[9]

The Palatine Society suffered a slightly less abrupt end. Work slowed following the sudden death in 1790 of its director, Johann Jakob Hemmer (1733–90), who had painstakingly edited the *Ephemerides*. The last volume, covering 1792, was issued in 1795, the year the French Army invaded Bavaria, occupied Mannheim, and destroyed Hemmer's files.[10]

That Hemmer and his collaborators never mapped their data reflects a narrow focus on observation, an excess of data, a dearth of hypotheses, and little or no appreciation of spatial patterns.[11] The typical observer was a cleric or physician

who hoped the project would contribute to a fuller understanding of the earthly effects of lunar occultations and other astronomical events. The *Ephemerides* are rich in tables reporting averages and extremes, and in time-series graphs allowing a visual comparison of various stations, with profiles aligned to the same horizontal axis, but no maps. Though Brandes might have attempted to collect the data himself, the convenience of a readily available, suitably comprehensive dataset no doubt fueled his curiosity.

To appreciate the geographic coverage of the Palatine network, I constructed figure 5.2 using tables in Kington's book, a 1985 article on the Palatine project by historian of science David Cassidy, and an 1826 article about barometric pressure fluctuations in Europe by John Frederic Daniell (1790–1845), an English chemist who had invented a barometer and authored a treatise on meteorology.[12] Cassidy and Kington listed the years during which the various Palatine observers contributed to the *Ephemerides*, and Daniell provided geographic coordinates and local details useful in correlating obsolete place-names with present-day locations. His Pyschminsk (in western Siberia) was apparently Cassidy's Pyshmen (in the Urals), missing entirely from Kington's list, as was Godthaab (today's Nuuk), on the west coast of Greenland. The map's upper panel shows the full extent of the Palatine network, including outliers in Pyschminsk, Godthaab, and Cambridge, Massachusetts. The lower panel shows a more compact portion within a few hundred miles of Mannheim yet sufficiently broad for a snapshot of a noteworthy weather system. A visual comparison of the solid dots in the lower panel with the posthumous reconstruction of Brandes's 1783 map (see figure 5.1) suggests that the collective effort of Palatine observers active that year was sufficient to reveal the low-pressure cell centered over the English Channel on March 6 but only partly surrounded by the map's isobars.

Although the network's outliers contributed nothing to Brandes's portrait of the March 6 storm, they imply that a wider network of correspondents could capture more revealing weather data. Had the Palatine network encompassed the British Isles and the Iberian Peninsula, Brandes might have more fully appreciated the generally west-to-east movement of weather systems, a directional bias that made timely synchronous weather maps a useful forecasting tool.[13] Indeed, the eastward movement of storm systems at least partly accounts for the success of meteorological networks established in the United States in the 1850s, when telegraph

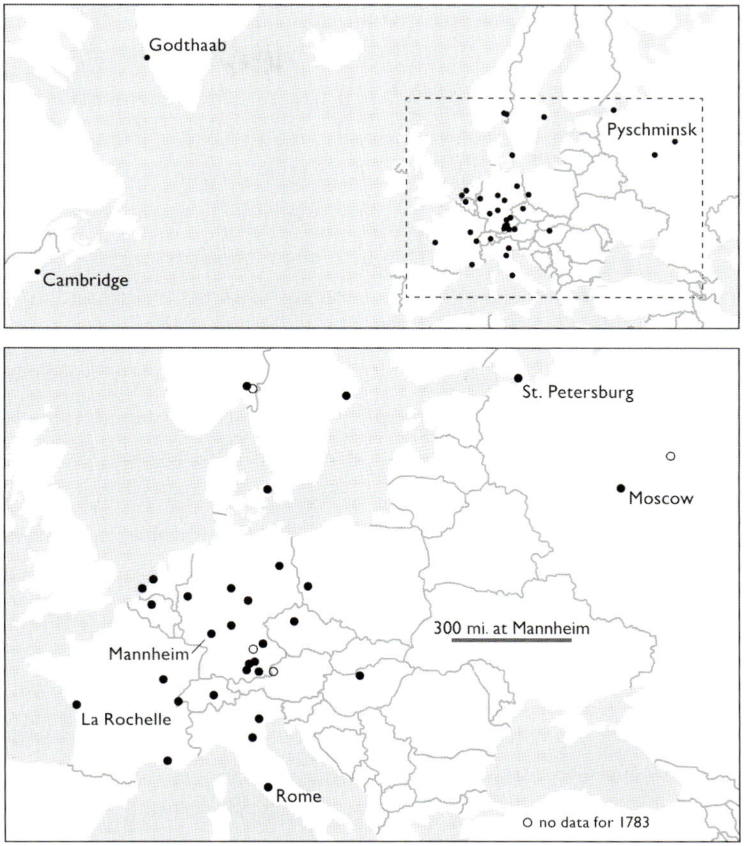

Figure 5.2. Locations of observers participating in the Palatine network. Open dots in the lower panel indicate stations not reporting during 1783. Present-day international boundaries provide a frame of reference. The dashed line on the upper map shows the area covered by the lower map. Because the maps are cast on a Mercator projection, on which scale increases with latitude, the scale bar refers specifically to the latitude of Mannheim.

wires reaching westward like tentacles strung along roads and railways promised weather forecasts beneficial to population centers east of the Appalachians.

Another physicist, Joseph Henry (1797–1878), was the primary proponent of the American project. In 1819, Henry enrolled as a student at the Albany Academy in upstate New York, where he presented his first scientific paper in 1824 and was appointed professor of mathematics and natural philosophy in 1826.[14] Henry was an early participant in the meteorological network launched

the previous year by New York surveyor-general Simeon DeWitt, who supplied the state's sixty-two private academies with standardized thermometers and rain gauges as well as instructions for taking daily measurements, ultimately summarized in a pioneering 1855 report on the state's climate.[15] DeWitt, who had championed the Erie Canal, was also head of the Board of Regents, which oversaw schools throughout the state.[16] This position empowered him to recruit observers, who had the satisfaction of participating in a major scientific endeavor while their employers shared in the state's "Literature Fund," which DeWitt controlled.

Henry is remembered for groundbreaking experiments in electricity and magnetism that made his last name the fundamental unit of electrical induction. These experiments included a telegraphic device, demonstrated in 1831, whereby an electric current traveled through a 1,060-ft. wire and rang a bell.[17] He was one of several experimenters to develop an electromagnetic relay able to transfer a signal from one telegraph circuit to another.[18] Relays could not only amplify a signal but also connect multiple lines together in a single long-distance circuit. Henry was less interested in the electric telegraph's commercial possibilities than Samuel F. B. Morse (1791–1872), a portrait painter whose experiments, patents, and business success benefited from Henry's discoveries.

In 1846, two years after Morse's pioneering forty-four-mile telegraphic transmission between Baltimore and Washington, the regents of the newly established Smithsonian Institution elected Henry secretary. In that capacity as the Smithsonian's CEO, he launched and nurtured a variety of projects in international scientific communication, terrestrial magnetism, agriculture, natural history, astronomy, and meteorology, to name a few, as well as a detailed census of the holdings of public and other institutional libraries.[19] Although Henry was well aware of the electric telegraph's ability to transmit synchronous weather data rapidly to Washington for a timely weather map, his first meteorological effort was a correspondence network, akin to the Palatine project.[20] As the state-level counts in figure 5.3 indicate, the Smithsonian's network expanded markedly between 1849, the first year its annual report included a list of observers and locations, and 1860, the year before the Civil War began to disrupt operations.

Expansion of the network was hardly straightforward—state tallies fluctuated from year to year because many volunteer observers participated for only a year or two. Comparison of the maps for 1849 and 1860 reflects efforts to extend the

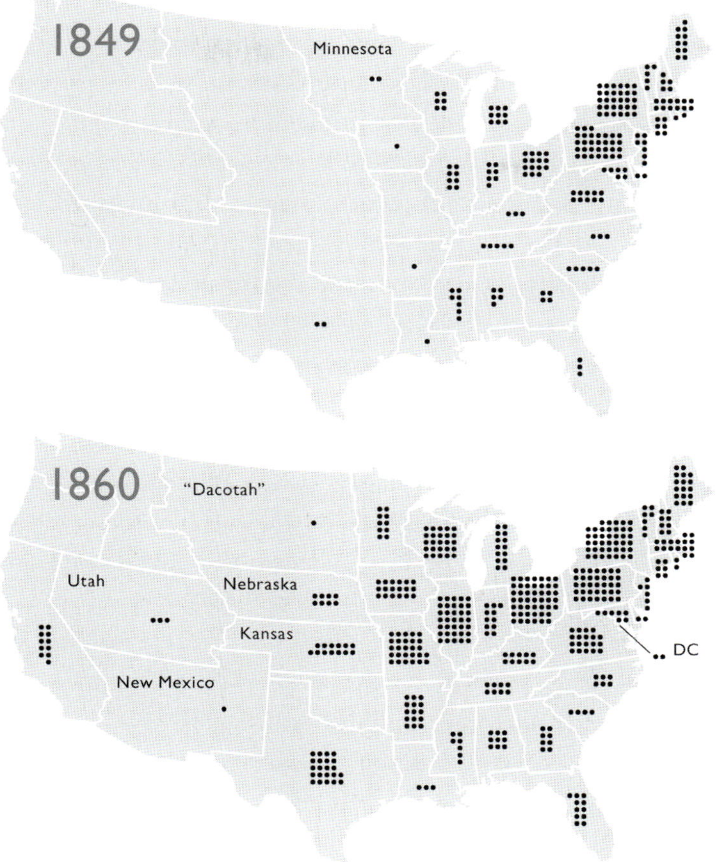

Figure 5.3. Number of meteorological observers reporting to the Smithsonian Institution, by state, in 1849 and 1860.

network to the west and south, to capture an ever more representative picture of atmospheric processes. Although prior annual reports hailed a steady net increase in the number of observers, the 1856 installment "regretted .... that the observers are not more uniformly distributed over the whole country." Singled out as "deficient" were Indiana, Kentucky, Tennessee, Mississippi, Arkansas, Louisiana, and Texas.[21] But by decade's end, observers were forwarding registers from more than a hundred locations between the Mississippi and the Pacific. Even so, for much of the country, the network was less dense than Henry's ambition that stations "be established not over sixty miles apart."[22]

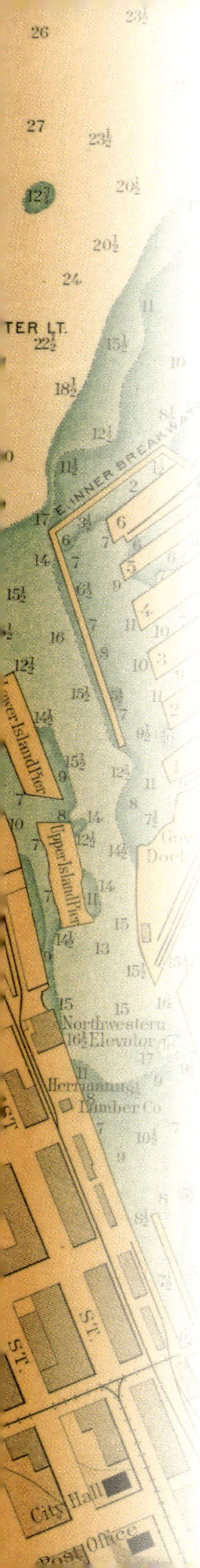

How less dense is suggested by the 56 × 86 cm (22.0 × 33.9 in.) “map showing the stations where meteorological observations are made in the United States under the direction of the Smithsonian Institution [in] 1851.”[23] A cartography reference specialist at the Library of Congress pointed it out when I asked about large basemaps produced by the government in the 1850s. To show details, I extracted and slightly enlarged the portion covering New York State (figure 5.4), which suggests that meteorological observers, presumably marked by small plus signs, were more than sixty miles apart. And as the maps of observer counts by state (see figure 5.3) indicate, New York was more heavily covered than most states. But as the clustering of plus signs along the St. Lawrence River and in a band reaching south from Oswego to the Pennsylvania border suggests, the dot-matrix symbols in figure 5.3 hide considerable spatial variation within the state. If Henry’s “sixty miles” is a meaningful threshold, this and other moderately populated parts of the country lacked adequate coverage of everyday weather and major storms.[24]

Despite its stated connection to the Smithsonian, the map hardly reflects Joseph Henry’s commitment to reliability and precision. Indeed, its lack of precision is an enigma. There is no mention of Joseph Henry inside or outside the neatline, and even though the map is attributed to the Smithsonian Institution, the Library of Congress has no record of a compiler, an engraver, a publisher, a printer, or even a former owner. I found no mention of anything like it in the Smithsonian’s annual reports, and nothing relevant in the Joseph Henry Papers, a carefully curated historical archive. In addition, the map lacks both a scale and a key. Moreover, my assumption that the plus signs represent meteorological observers is only partly confirmed by the Smithsonian’s annual reports for 1851 and 1852, which list observers by name and place of residence.[25] Only sixteen of the twenty-six places marked with plus signs are on the list of New York observers for 1851, four of the state’s listed observers are not marked by a plus sign, and three places with an observer are not even on the map.[26] For 1852, the corresponding counts are fourteen, one, and five.

One of these missing places is Syracuse, a thriving Erie Canal port with an observer listed for both years. Was it suppressed to avoid graphic conflict with the *Y* in the large New York label? Not likely, given the mapmaker’s flagrant disregard for locational accuracy. Among other examples, Rochester and Rome are noticeably west of their true positions, Oswego has been displaced to the northeast, and

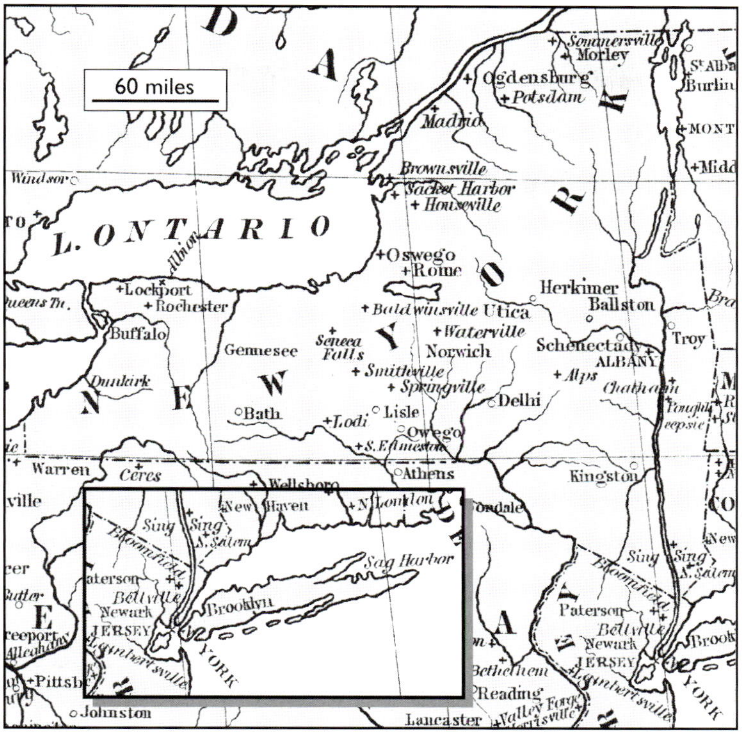

Figure 5.4. Portion of the 1851 Smithsonian observers map covering New York State.

South Edmeston is shown southwest of Oswego when it should be northeast. Another flip-flop is apparent along the Hudson River: although Kingston and Poughkeepsie are on their respective sides of the river, Poughkeepsie is shown significantly north of Kingston when it is actually farther to the south. Outrageous errors in neighboring states are apparent for Montpelier, Vermont, and for Hartford and New Haven, Connecticut. Although this level of brazen sloppiness is difficult to explain, a naïve but technically competent volunteer engraver might have intended to promote the Smithsonian project by calling attention to both the network's reach and the need to improve coverage. Who knows?

Paying observers in underrepresented regions could have helped Henry expand and densify the network. As he noted in the 1859 report, "Frequent applications are made to the Institution offering to record meteorological observations for a reasonable pecuniary compensation"—some potential collaborators wouldn't collaborate unless they were paid.[27] Even if funds were available, Henry believed the "difficulties of discrimination [were] altogether disproportionate to the advantage gained"—he feared the resentment and injured feelings of unpaid observers. Although the publications he occasionally gave volunteers were "of some value," their primary compensation was "the pleasure to be derived from the study of the changes of the phenomena of nature [as well as] the consciousness of cooperating with many others in a system intended to advance an important branch of science."[28] Nicely put. Indeed, some volunteers even purchased their own instruments from an approved manufacturer.[29]

In addition to calibrated instruments, observers received blank sheets for recording observations taken daily at 7:00 a.m., 2:00 p.m., and 9:00 p.m. and mailed back monthly.[30] Several hundred observers inundated the Smithsonian with data that required a staff of "twelve to fifteen persons, many of them female," to reduce barometric pressure measurements to sea level to remove the influence of elevation, as required for mapping; add the new readings to a growing pen-and-paper database; and calculate summary statistics essential for a national climatology.[31] Clerical staff and postal expenses were subsidized by the Patent Office, where commissioner Charles Mason (1804–82) viewed Henry's meteorology project as a valuable contribution to both agriculture and general knowledge.[32]

Henry enjoyed the cooperation of other government agencies. In 1852, the New York State Regents reestablished their observer network and provided instruments to a limited number of participants, and Massachusetts did likewise.[33] Several annual reports list members of these participating networks separately, apart from Smithsonian volunteers living in the same state.[34] In addition, the Smithsonian benefited from observations at military posts, arranged by the army's surgeon general, who was particularly interested in weather extremes. Some annual reports included a list of army posts that contributed to the Smithsonian database.[35] In addition, employees of the federal Light-House Board reported conditions on the Atlantic and Pacific coasts, and members of the Army Corps of Topographical Engineers contributed observations at stations on the Great Lakes.[36]

Reliance on other agencies is inherently fragile. In 1857, Mason retired as commissioner of patents, and both his successor and his successor's successor lacked his appreciation of meteorological intelligence.[37] In the report for 1859, Henry boasted "the observers have very much increased in number," but noted warily that Patent Office support was "less than in previous years."[38] A year later, the congressional appropriation for meteorology, funneled through the Patent Office, was "suddenly and unexpectedly suspended."[39] Forced to cut back, Henry continued to collect data but was "obliged to stop the reductions"—the adjustments to sea level that provided comparable pressure readings for mapping—which he hoped would "be resumed again before the observations accumulated to an unwieldy bulk." Computations could be deferred while network operations remained intact.

Henry was more interested in timely data captured only once a day by a sparse network of telegraph operators who replaced their usual "I'm here" message sent out when they arrived at work with a brief report on early morning weather and wind direction.[40] Although his telegraphic network, initiated in 1856, was receiving reports from only thirty-two stations by the end of 1858, and only forty-five stations by 1860, Henry eagerly shared the information on a map described in his 1858 report.

> An object of much interest at the Smithsonian building is a daily exhibition on a large map of the condition of the weather over a considerable portion of the United States. The reports are received about ten o'clock in the morning, and the changes on the maps are made by temporarily attaching to the several stations pieces of card of different colors to denote different conditions of the weather as to clearness, cloudiness, rain, or snow. This map is not only of interest to visitors in exhibiting the kind of weather which their friends at a distance are experiencing, but is also of importance in determining at a glance the probable changes which may soon be expected. [41]

The map caught the fancy of congressmen who stopped by regularly for his best-guess forecasts, based on the generally reliable west-to-east progression of storms.

Henry provided more details in his 1859 report.

> A large map is pasted on a wooden surface, into which, at each station of observation, a pin is inserted, to which a card can be temporarily attached. The observations are made at about seven o'clock in the morning, and as soon as the

> results are received at the Institution, an assistant attaches a card, indicating the kind of weather at the time—rain being indicated by a black card, cloudiness by a brown one, snow by a blue one, and clear sky by a white card.[42]

Particularly ominous was a black card posted in the morning for Cincinnati, indicating that

> .... a rainstorm may confidently be expected at Washington at about seven o'clock in the evening. Indeed, so uniformly has this prediction been verified, that last winter the advertising in the afternoon papers of the lectures to be delivered at the Institution that evening was governed by the condition of the weather in the morning at Cincinnati—a rainy morning at the latter inducing a postponement of the lecture.[43]

Circular cards with an arrow painted at the center included eight holes punched around the edge so that they could be hung to show prevailing wind direction.[44] The cards were approximately an inch in diameter.[45]

Aside from Henry's description of the map as "large," we know little about its overall size, which I assume to be at least 3 ft. wide, typical of mid-nineteenth-century outline maps of the United States. (Inquiries about likely sources turned up the map in figure 5.4.) Surely, the map was larger than the version in the 1933 painting by Louise Rochon Hoover. Commissioned by the Smithsonian and titled *Professor Henry Posts Daily Weather Map in Smithsonian Institution Building, 1858*, the painting shows a stern, somewhat pensive Henry addressing a well-to-do couple in period dress on opposite sides of a map barely two feet wide and less detailed than the curtains, fluted columns, display table, and ornate chair in the background.[46] The painting is posted on the Smithsonian website, where a zoomed-in view shows Henry sticking a small square card on central Texas, not part of his telegraphic network. I saw no point in reproducing it here, and it's easy to find online.

Hoover's flawed reconstruction is the only graphic example I could find. Smithsonian annual reports of the era were rarely illustrated, and whatever examples or paraphernalia might have been preserved in the National Archives were probably lost in the January 1865 fire that destroyed much of the meteorology project's records and equipment.[47] Historian of science James Fleming, who wrote the definitive history of American weather science prior to 1870, discovered plans to resume telegraphic reports and add a second map at the Capitol but found no evidence of these displays.

For a sense of the Smithsonian map's content, I plotted the telegraphic stations listed in its annual reports for 1858, 1859, and 1860 and added telegraph lines either mentioned in these reports or shown on Charles Barr's 1853 map *Telegraph Stations in the United States, the Canadas & Nova Scotia*, in the Library of Congress.[48] In 1857, Henry applauded "the National Telegraph [Company] line for a series of observations from New Orleans to New York, and as far westward as Cincinnati, Ohio."[49] Although the 1853 map does not name individual companies, it describes an unambiguous route from New Orleans to New York as well as connections between Washington and Parkersburg, more than halfway to Cincinnati, that are more straightforward than the route westward, complicated by multiple pathways.[50] Figure 5.5 addresses this relative ambiguity with solid or dashed lines representing greater or lesser certainty in their location.[51] Dashed lines also connect with the few cities that aren't part of Barr's 1853 network.

Dots that vary in size or grayscale reflect the mostly progressive growth of the Smithsonian telegraph network. The thirty-one large black dots for stations contributing during all three years describe the system's early and more durable reach northeastward to New York, westward to Cincinnati, and southwestward to New Orleans, along a route that approximates the physiographic boundary between the Piedmont and the comparatively level coastal plain. The minimalist network in the South is underscored by lateral branches to Lynchburg and Norfolk, Virginia; Wilmington, North Carolina; Charleston, South Carolina; and Savannah and Atlanta, Georgia. The long Shenandoah Valley, west of the Blue Ridge Mountains, provided an efficient route to Knoxville, Tennessee, not included on the 1853 Barr map. The ten dots in the second-most common category show significant extensions beyond New York to Burlington, Vermont; beyond Cleveland to northern Illinois and Iowa; and beyond Cincinnati to St. Louis, Missouri, and Cairo, Illinois.

Henry's network reflects dominant atmospheric currents as well as the geography of telegraph lines. Although the telegraph network was comparatively well developed in New York and New England, the weather there was less relevant to

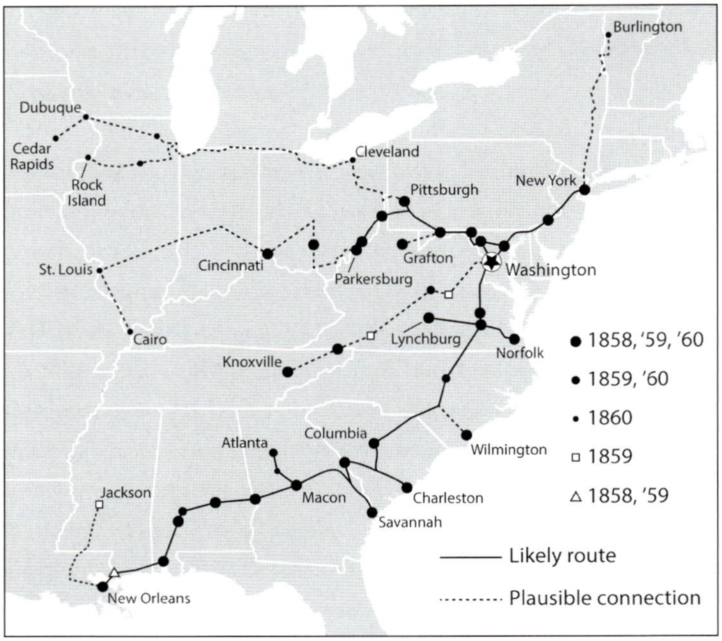

Figure 5.5. Stations from which telegraphic reports of weather were received at the Smithsonian Institution any time between 1858 and 1860.

the Washington forecast than conditions in the Midwest. The handful of temporary additions such as Jackson, Mississippi, and two stations in central and southwestern Virginia that contributed reports only in 1859 were apparently of limited interest in Washington. That Gainesville, Mississippi (near the Louisiana border and the Gulf of Mexico), dropped out after reporting during 1858 and 1859 was probably of little consequence because of its proximity to New Orleans.

Particularly noteworthy is the elongated cluster of nine dots to the west and northwest between Washington and Cincinnati and including Pittsburgh. Their importance to Henry and his congressional supporters is apparent in the 1860 report, in which he expressed "regret that frequent intermissions take place in the receipt of the telegrams from places directly west of the city of Washington, especially as we are more immediately interested in these, since they afford the means of predicting with considerable certainty the character of the weather sometimes a day or more in advance."[52] The Washington establishment was a key part of Henry's consumer base.

Hiccups in the 1860 telegraphic reports were minor compared with the disruption after South Carolina seceded from the Union in late December 1860 and ten more states followed over the next six months. In his annual report for 1861, Henry lamented, "The system of meteorological observations, organized and carried on successfully for several years past, has suffered more from the disturbed condition of the country than any other part of the operations of the Smithsonian establishment."[53] The Civil War affected the correspondence network when federal troops were withdrawn from military posts in the nation's interior and along the Pacific coast, but the war's impact on the telegraph network was even greater. Not only were reports from the South "interrupted" but "the wires from the North and West [were] so entirely occupied by public business that no use of them could be obtained for scientific purposes."[54]

Henry's network never fully recovered after the war ended in 1865. Congress no longer funded the reduction of weather data and no longer allowed correspondents to send in their monthly registers postage free.[55] Weather telegraphy had partly resumed in 1862, with reports added that December from California and the Rocky Mountains, but Henry judged the data "not sufficient to enable us to predict, with much probability, the changes of weather, without the additional information from the South and Southwest."[56] Although the war's end might have restored telegraphic reports from these regions, repairing the damage caused by the 1865 fire diverted funds from meteorology and other projects, and telegraph companies were less willing to provide free use of their lines and operators.[57] Henry pressed Congress for funds to revive the meteorology project, but his proposal was less persuasive than an alternative plan introduced by Wisconsin Congressman Halbert E. Paine (1826–1905), who favored a national weather service run by the military.[58] Paine's proposal, which became law in February 9, 1870, created a weather service in the Army Signal Corps, which had provided both visual and electronic long-distance communications during the war and needed a new mission.[59]

Two men guided the new weather service during its first decade: Albert J. Myer (1828–80) and Cleveland Abbe (1838–1916). Myer, who had begun his military career as an army surgeon and written a textbook for military officers on semaphore signaling, commanded the Signal Corps during the Civil War.[60] In 1870, as a brigadier general and the army's chief signal officer, he dispatched

military weather observers (mostly sergeants) to selected cities to set up telegraphic weather stations. In early 1871, Myer hired Abbe, a civilian, as his assistant in charge of weather forecasting. In 1868, Abbe had been appointed director of the privately funded Cincinnati Observatory, and in September 1869 he began using telegraphic weather reports from a small network of observers to produce a weather map and predictions for the Cincinnati Chamber of Commerce—commodities speculators are keenly interested in weather forecasts.[61]

Abbe's experiment was so well received that the telegraph company decided to share the recognition. In February 1870, the manager of the local Western Union office, through which Abbe received his data, applied the company's brand to a similar map reproduced using tissue-thin outline maps interleaved with carbon paper. The Library of Congress has one of these outline maps (figure 5.6), which illustrates Western Union's appropriation of Abbe's project as well as the symbols used to report weather conditions, wind direction, and temperature (but not pressure) at over two dozen stations reaching from Cheyenne, Wyoming, to Havana, Cuba, to Halifax, Nova Scotia. Focused on observations at stations, the maps had point symbols but no isolines. In May 1870, Abbe ceded his project to Western Union, which phased out the Cincinnati weather map that October in anticipation of the Signal Office start-up.

Few civilians were as well prepared as Abbe to help launch the new national weather service. Trained in astronomy and meteorology, he was not only the Signal Service's lead forecaster for many years but also its de facto chief scientist through 1891, when weather forecasting was transferred to the US Weather Bureau, a civilian agency in the Department of Agriculture. Until his death in 1916, Abbe continued to work as a government meteorologist, publishing scientific papers that strengthened the field's theoretical foundations.[62]

To provide a fully synchronous weather map, Signal Service observers took measurements at 7:35 a.m., 4:35 p.m., and 11:35 p.m., Washington time— observers on the West Coast were apparently conditioned, as soldiers, to rise well before dawn. They then converted their observations into words representing the station name, date, time, barometric pressure, temperature, humidity, weather condition (heavy rain, snow, and so on), direction and velocity of wind, variability and pressure of wind, upper and lower cloud conditions, rainfall since the last report (to the nearest hundredth inch), and where relevant, the rise or fall (in inches) of an

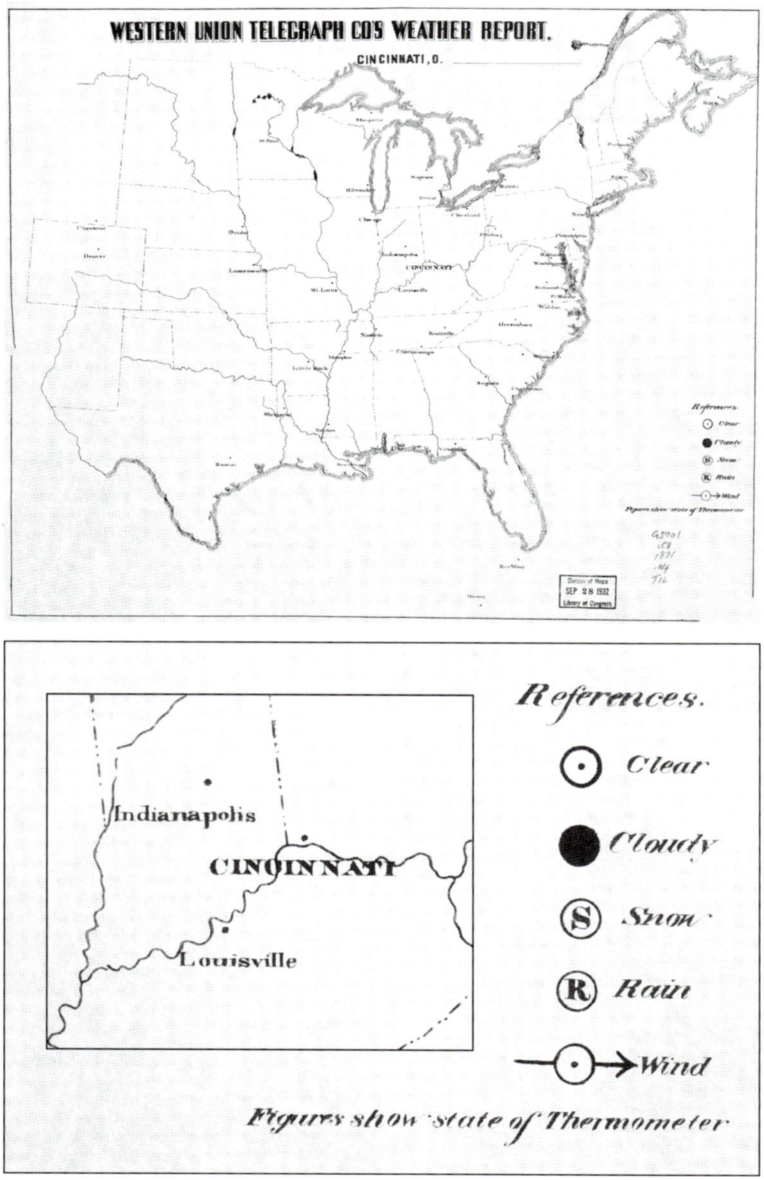

Figure 5.6. *Top:* Blank-outline weather map used by Western Union's Cincinnati office, ca. 1870. Geography and Map Division, Library of Congress. 32 × 39 cm. *Bottom*: Full-size excerpts of the map key and the area around Cincinnati.

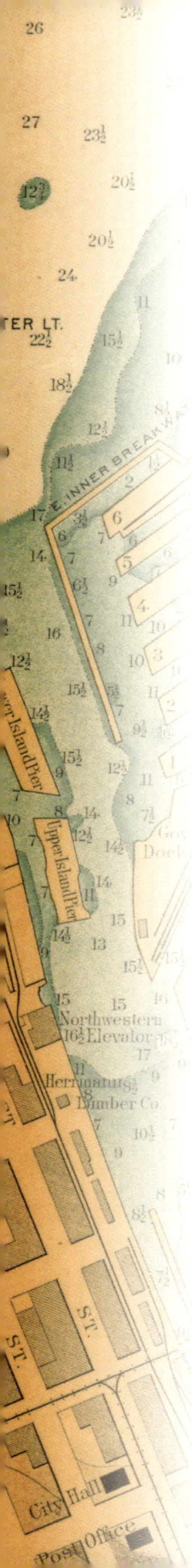

adjacent river.[63] On the sixth day of the month, for instance, time was coded *Cat*, *Gave*, and *Ned*, respectively, to represent the morning, afternoon, and late-evening reports; barometer readings of 28.63, 28.64, and 28.65 in. became *Falstaff*, *Falter*, and *Familiar*; and when the wind was blowing from the northeast, *Judith* meant heavy snow and *Fanny* meant foggy. Cipher codes such as these compressed the characters that were telegraphed by a factor of two or more, and so reduced transmission time and cost accordingly.[64] Each morning report consisted of twenty words, whereas the afternoon and night reports included only ten words.

Reporting was organized as a network of "circuits," each represented in figure 5.7 by a thick gray line and consisting of two or more stations that simultaneously received whatever any other station keyed into the same circuit. Some cities, such as Keokuk, Iowa, and Leavenworth, Kansas, belonged to just one circuit, whereas St. Louis belonged to two: the two-station circuit it shared with Leavenworth and the three-station circuit that included Chicago and Cairo, Illinois. Although data entered at Leavenworth could be keyed into the Cairo–St. Louis–Chicago circuit at St. Louis, the transfer was not automatic, and the station at Leavenworth was not inherently privy to data entered at Cairo. Moreover, although the wire connecting Cairo and Chicago might have passed through St. Louis, a city's intermediate position on a circuit list did not necessarily mean that it was electronically intermediate. For example, although the list of stations on the circuit between New York and Washington places Cape May between Philadelphia and Baltimore, the wire from Philadelphia to Baltimore most certainly did not pass through Cape May.

Figure 5.7 is my attempt to describe operations in June 1871, eight months after the system's start-up. Although I tried to mimic the network's geography, the limited width of a book page required bending downward the westward tentacle to the trio of stations on the Pacific coast. After a little experimentation, I was able to fit the twenty-five-circuit network into a squarish space that accommodates relatively long and complicated circuits, including the eleven-station circuit linking New York with Lake City, Florida, about 100 miles west of Jacksonville, as well as the so-called "special circuits" linking Washington with New York, Washington with New Orleans, and Washington with Chicago. *Special* indicates an exclusive circuit encompassing just a pair of cities also linked to each other by another circuit shared with other cities.

Figure 5.7. Schematic map of the twenty-five telegraphic circuits operating in June 1871. Place-names reflect the original spelling in the Signal Service report for 1871, including, for example, Lynchburgh and Du Luth.

A few prominent circuits dominate the map: the 2,337-mi. branch from Chicago to San Francisco, with extensions up and down the Pacific coast; a north–south circuit down the middle of the country from Chicago to New Orleans; the aforementioned East Coast circuit from New York to Florida; a pair of circuits capturing conditions along the Gulf Coast at Galveston, Texas, New Orleans, and Mobile, Alabama; and the 1,019-mi. circuit encompassing Milwaukee, Wisconsin, New York City, and several ports along the southern shore of the lower Great Lakes.[65] A dashed-line rectangle highlighting Chicago's prominence as the station on more circuits than any other station partly obscures the Windy City's intermediate position on the Milwaukee–New York circuit, which captured conditions upwind from the country's economic core. Although the Chicago office

became a district forecast center in 1894, in the 1870s its soldier observers mostly read instruments and forwarded data.[66]

Precise schedules listed each circuit's stations along with their prescribed "local time" for sending morning, afternoon, and nighttime reports.[67] Some circuits, such as the link between Boston and Mount Washington, New Hampshire, the highest peak east of the Mississippi, called for transmission in only one direction, but most circuits served noteworthy populations at both ends. For example, the morning report from Burlington, Vermont, entered at 8:15 a.m. local time on the circuit with Boston, was followed at 8:45 a.m. by a report from Boston that included information from New York, Portland, Maine, Buffalo, and Oswego: five cities apparently deemed sufficient to warn of imminent storms from the west and south.[68] Because Burlington is roughly 2.15° west of Boston, the local time in Vermont was only 8:36 a.m.[69]

By contrast, Oswego, New York, on the opposite side of the Adirondacks from Burlington, received 53 reports, just two fewer than the Signal Service forecast center in Washington.[70] Although this apparent imbalance might be attributed to Oswego's position as a Great Lakes port on an important branch of the Erie Canal, a more convincing explanation lies in its participation in the circuit between Milwaukee and New York City, and the funneling of reports from the Southeast and the West through Chicago, which also received 53 reports.

Although data moved quickly through the system, the sequential keying of observations into the network mirrored not only the west-to-east movement of storms but also the flow of moisture northward from the Gulf of Mexico. Scheduling tables show that reports from Norfolk, and Key West and Jacksonville, Florida, were transferred to the Augusta–Mobile–New Orleans circuit before being sent north to Chicago, where they were aggregated with data from San Francisco, the Midwest, and Pittsburgh and transferred to the Milwaukee–New York circuit, on which they could be read at Oswego.[71] Although Washington took reports from Norfolk, Lake City, and intervening stations on the Lake City–New York circuit, data from Knoxville, Baltimore, and points north reached Chicago through New York City, where they were placed on the Milwaukee–New York circuit along with reports from New England. Scheduling tables were silent about the role of the network's three "special circuits," which could have supported strategic communications between meteorologists at important regional hubs or provided backup when key telegraph lines were down.

The 1871 Signal Service report mentioned three additional circuits (not included in figure 5.7): one international and two on the Western frontier. Montreal's McGill Observatory, which telegraphed observations three times a day (except on Sundays) to Portland, Maine, had access to all reports available there.[72] Although scheduling tables suggest that the Montreal data might not have been distributed electronically beyond New York, the observatory also contributed climate data to the larger postal network. A second supplementary circuit connected to the San Francisco line at Corinne, Utah, included Virginia City, Nevada, to the south and Fort Benton, in Montana Territory, to the north. A third supplementary circuit connecting at Cheyenne, Wyoming, included Denver, Colorado, and Santa Fe, New Mexico, both to the south.[73] Although reports on these circuits might not have traveled telegraphically beyond Chicago, the connections no doubt helped the Signal Service inform Washington about Indian unrest on the frontier.[74]

Meteorological surveillance relied on military discipline not only to take each station's thrice-daily readings and transmit them on schedule but also to enrich the nation's climate database with additional readings taken daily at 7:00 a.m., 2:00 p.m., and 9:00 p.m., local time, and recorded on a form mailed weekly to the central office.[75] In addition, the barometer was read daily at midnight, Washington time, and if a station's pressure had risen or fallen by more than 0.15 in. since the morning reading, the new measurement was immediately reported by telegraph to headquarters, along with weather conditions and the direction and velocity of wind.[76] The central office needed to know of changes that might reflect the emergence of a new pressure cell as well as significant changes in the movement of existing lows and highs.

Formal rules governed station operation and the flow of information. Instead of merely delivering a report to the local telegraph operator, the observer had to remain at the telegraph office until the data were sent and the designated observations were received from other stations. While waiting, the observer could begin decoding ciphers from distant stations, composing a bulletin for posting at the local board of trade or chamber of commerce or sketching a rough draft of a weather map.[77]

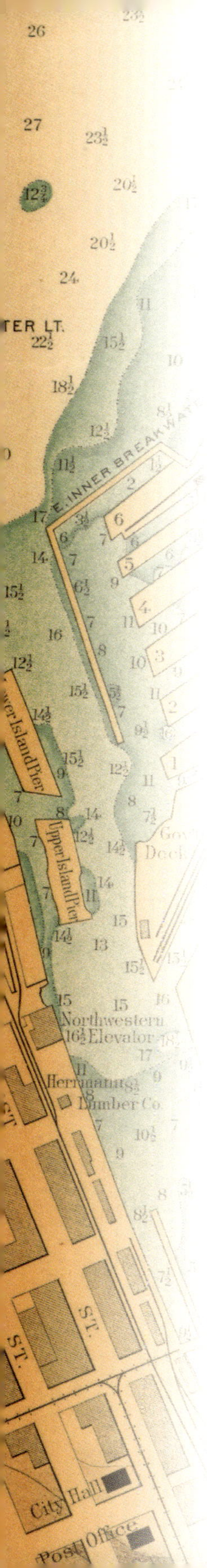

Some stations had a small printing press for reproducing maps to be distributed to local newspapers and, if time permitted, to individuals who requested one.[78] Others had a so-called "manifold press," which used carbon paper, preprinted outline maps (e.g., figure 5.6, top), metal stamps, and a mallet for making ten to fifteen copies at a time. Carbon sheets were interleaved between preprinted outline maps, and metal stamps were used to add wind direction and state of weather for each station shown. Guidelines told the observer/mapmaker to "hold the die perpendicular to the map and strike it hard enough to make the impression go through all the maps without cutting the upper one."[79] Curiously optimistic, the instructions added, "The actual amount of force required can only be learned by practice." Observations for individual stations were also posted on a large outline map labeled *War Department Weather Map* using arrows and colored disks.[80]

Posting observations was fine, but the central office kept a tight rein on forecasting. "In furnishing weather reports and items for publication," one instruction warned, "observers must confine themselves strictly to the instructions issued from this office, and will not, under any circumstances, publish, or cause to be published, forecasts or predictions of the weather not originated by this office."[81] Rogue forecasts could undermine the weather service's reputation as well as its command-and-control structure. As weather prediction and its limits became better understood and more reliable, the central office delegated some forecasting responsibilities to district forecasters and local offices.

Contracts with Western Union and other companies guaranteed the dedicated use of key parts of the nation's telegraph network during fixed parts of the day. Although this system covered most of the country, government-owned lines were needed to reach remote areas in the interior and along thinly settled parts of the coast. Although commercial lines connected observers in Key West and Punta Rassa, Florida, with Washington, an independent military telegraph network delivered forecasts to outposts in the interior, and a separate "sea-coast telegraph" system allowed the timely posting of storm warnings at lighthouses and other strategic points along the Atlantic coast.[82] A short history of the Signal Service published in 1884 included a map (figure 5.8) that highlighted government-run telegraph lines along three sections of the coast between Smithville, North Carolina, and Sandy Hook, New Jersey. Connected to the central forecasting office in Washington, these circuits also collected routine information on winds, waves, and tides, and informed nearby lifesaving stations about reported shipwrecks. The Sea-Coast Telegraph reflects a mature network that had expanded to disseminate as well as capture weather data.

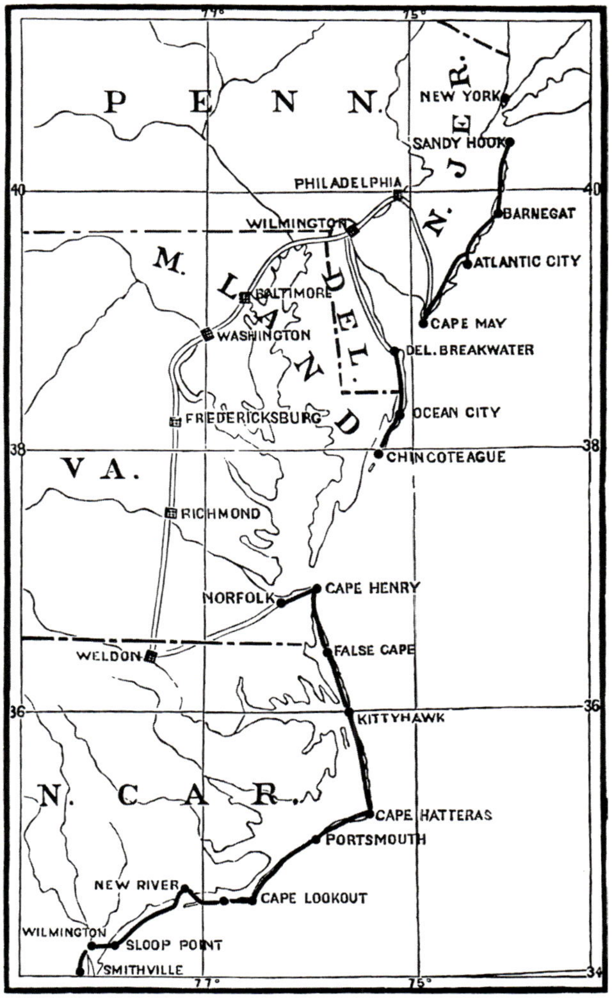

Figure 5.8. The Sea-Coast Telegraph, ca. 1884. The white (double-line) symbols show connections of black coastal telegraph lines to Washington.

Operational meteorology followed the railroads and telegraph companies in adopting standard time, which was widespread by the mid-1880s.[83] When the Signal Service began operations on November 1, 1870, synchronous measurements were tied to "Washington time," which was local time at 77°3' W. On January 1, 1887, Washington time gave way to 75th meridian time, the standard for the easternmost of the country's time zones. Meanwhile, self-recording instruments that captured hourly data had obviated separate readings for the climate database. On July 1, 1888, improved instruments, a fuller understanding of storms, and a denser observation network culminated in the replacement of thrice-daily measurements with a uniform system of climate and telegraph readings taken at 8:00 a.m. and 8:00 p.m.[84]

In 1891, operational meteorology was transferred to the US Weather Bureau, a new agency set up within the Department of Agriculture to better address the needs of agriculture, commerce, and transportation. The Signal Service already had numerous nonmilitary scientists and clerks on its payroll, and noncommissioned officers who had received formal training in meteorology became civilian employees in the new agency. Although the number of stations had increased from fifty-five in 1871 to seventy-three in 1891, the system of telegraph circuits, though more geographically extensive, was largely unchanged.[85] In a 1922 study of Weather Bureau operations, Gustavus Weber, a policy analyst at the Institute for Government Research (a forerunner of the Brookings Institution), described a system of circuits functionally similar to the circuits "inaugurated at the very beginning of the Weather Service."[86] By then, 175 of the Bureau's two hundred "regular stations" were sending observations twice daily on a network of twenty-one circuits, and 140 stations were receiving data "as is adequate to their needs."[87]

These needs included making maps for the local forecaster, who depended on a snapshot of the most recent state of the atmosphere, as well as local newspapers that had arranged to print a weather map in their next edition. Wire reports told the mapmaker which weather condition symbols to use for each station shown and in which direction to point its wind arrow. Training and experience guided the sketching of isobars and isotherms. Although station maps drawn the same day in different cities were never identical, they were rarely radically different insofar as regulations called for mailing copies to Washington for review.

In the early 1880s, the Signal Service experimented with telegraphing instructions to a map drafter in New York, who would then replicate isolines drawn on a master map in Washington.[88] A grid was placed over a map at the central office, and each isotherm and isobar was coded as a chain of grid cells. Because the grid proved too coarse to guarantee an exact duplicate, mapmakers at the New York office soon returned to drawing their own isolines.

Newspaper weather maps caught the fancy of Harvard geography professor Robert DeCourcy Ward, who chronicled the genre's rise and current state in an 1894 *American Meteorological Journal* article that included examples from newspapers in Boston, Cincinnati, New Orleans, and San Francisco.[89] Figure 5.9 shows how the *San Francisco Examiner* described weather in the western third of the nation at 5:00 p.m. (Pacific time) on a day in late February when cold temperatures and clear skies accompanied high pressure over Montana. The map was drawn in India ink at the weather office in the early evening and taken to the newspaper, where a photoengraver reduced it to a smaller size and produced a zinc plate used in printing the next day's edition. Although drafting and image-transfer processes differed, the four maps cited by Ward used identical point symbols for weather conditions, similar lines for pressure (solid) and temperature (dashed or dotted), and prominent (larger, darker) letters to identify cells of high and low pressure. All were hand-lettered.

Despite similar symbols, Ward's four examples reflected different basemaps. Figure 5.10 highlights these disparities with distinctive lines that encircle the cities included on each map. As the dashed line on the left side of the map shows, none of the cities on the *Examiner*'s map were east of the Rocky Mountains, and cities in the West were conspicuously absent from the other three maps. Although the fourteen black dots representing cities common to the other three maps seem more or less broadly distributed throughout the Midwest and the South, regional biases are apparent in the *Boston Herald*'s favoring of cities in the Northeast, in the *New Orleans Times-Democrat*'s including all but five cities south of the Mason-Dixon Line, and in the *Cincinnati Tribune*'s collective coverage of the Midwest and Great Plains and its avoidance of cities in a broad belt between Wilmington, North Carolina, and Amarillo, Texas. Beyond speculation about motives or perceived importance, figure 5.10 demonstrates pointedly that the maps drawn in these four cities were hardly identical.

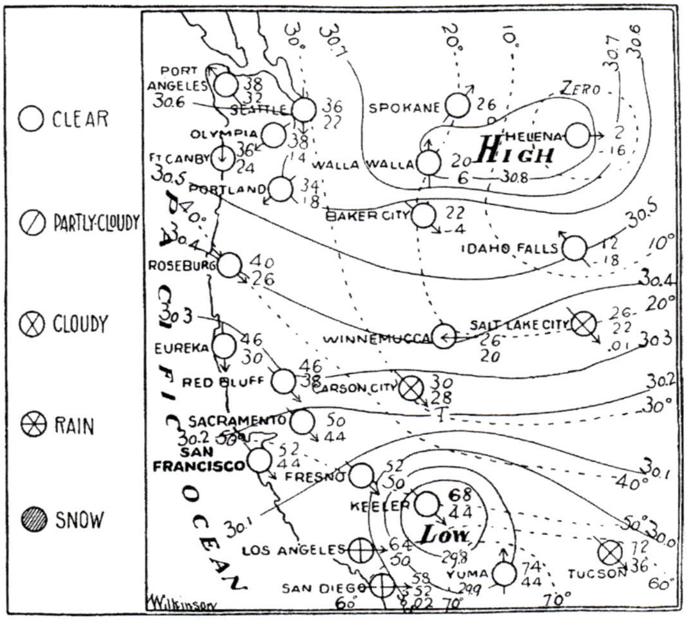

Figure 5.9. The *San Francisco Examiner*'s weather map for 5 p.m., Pacific time, February 21, 1894, was one of four examples in Robert DeCourcy Ward's essay on newspaper weather maps.

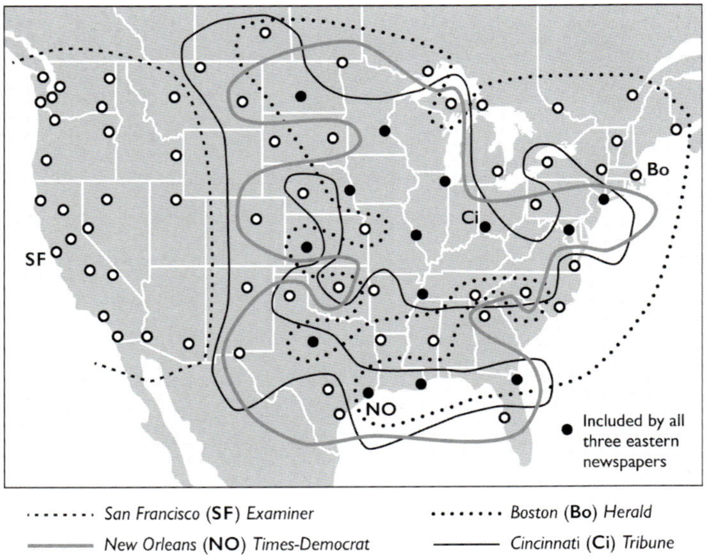

Figure 5.10. Differences in geographic coverage as defined by cities shown on the four example weather maps in Robert DeCourcy Ward's 1894 essay on newspaper weather maps.

In 1912, the Weather Bureau convinced 147 newspapers in ninety-one cities to run a daily weather map, plotted locally on a standard basemap with the prominent two-line title "U.S. Department of Agriculture"/"WEATHER BUREAU" at the top.[90] The Bureau's motives included promoting its brand and reducing the number of stations printing their own "station weather map" from 110 to only fifty-nine—printing, posting, and mailing weather maps had become a significant part of the daily routine at most stations, some of which employed full-time press operators. Offloading printing and distribution onto a local newspaper not only reduced costs but also greatly increased the number of map views if one counts the cooperating newspapers' combined daily circulation of 2,004,704.[91] And the local paper acquired a new and fascinating feature at little or no added cost.

Many of the 147 newspapers dropped the feature in the years following, but a revival of sorts occurred in 1935, when the Associated Press (AP) launched its AP Wirephoto network and began offering subscribing newspapers morning and afternoon weather maps along with a steady stream of news photographs.[92] An AP artist in Washington redrew an official Weather Bureau map with the coarser lines and labels able to survive the graphic degradation of scanners and printers that were little more than low-resolution fax machines. Replicated exactly in multiple cities, the Wirephoto weather map was smaller than its central office counterpart and showed fewer stations. In 1948, the Weather Bureau began using facsimile transmission to obviate redrawing the Washington map at individual weather stations.[93]

Industry dynamics, weather science, and electronic computing reshaped the newspaper weather map many times. Other news syndicates began to compete with Wirephoto, image transmission technology improved, fronts displaced isobars, short-term forecast maps superseded stale snapshots of recent conditions, and in the 1980s a revolution in newspaper graphics epitomized by *USA Today* replaced mundane black-and-white maps of pressure cells with full-color portraits of temperature zones.[94] Most newspapers now get their weather graphics over the internet from a specialized weather service such as AccuWeather®.

As ever-changing snapshots of the atmosphere, weather maps are clearly the single most numerous cartographic genre in the public sphere: more numerous than printed roadmaps, more numerous than topographic maps, and more numerous than maps in textbooks, newspapers, and magazines. More significant is their role in

operational forecasting and atmospheric research as an analytical tool, epitomized by the online "weather wall" with dynamic links to a multitude of high-resolution weather graphics, computer models, and radar and satellite images.[95] Maps are undeniably ubiquitous in weather science, where they are a basic tool of visual analysis.

Visualization is not a new role: in the late nineteenth and early twentieth centuries, the weather map was an indispensable training aid for would-be forecasters as well as experienced forecasters eager to improve their skill. Large (22 × 16 in.) versions of the morning and evening weather maps were lithographed in black ink on outline maps with shorelines, boundaries, and place-names in blue and distributed to weather stations, libraries, and individual subscribers.[96] The central office and many stations kept bound copies on file for self-study. George Bliss, forecaster in charge of the Philadelphia station, praised this practice in a 1917 *Scientific American* article.

> The most profound students of atmospheric physics .... were far from being the best forecasters. .... In order to excel in the profession, one must possess a special faculty for intuitively and quickly weighing the forces indicated on the weather map and calculating the resultant. This special faculty is developed by long and continued study and association with the maps, rather than by a profound study of atmospheric physics.[97]

Maps were also a standard feature in the *Monthly Weather Review*, an in-house periodical in which maps and tables summarized climate for the month or year and short articles discussed recent severe storms, forecasting practices, and new or improved instruments.

Among weather telegraphy's more intriguing summary graphics are the storm track maps of Frank Bigelow (1851–1924), a mathematician and former Episcopal minister who was a professor of meteorology (a research position) at the Weather Bureau between 1891 and 1910.[98] Best known for his work on the theory of atmospheric circulation, Bigelow illustrated his 1897 report *Storms, Storm Tracks, and Weather Forecasting* with a dozen maps, such as figure 5.11, which describes the eastward drift of storms formed in January, mostly in the Northwest or Texas.[99] To help forecasters understand seasonal and geographic trends, he tracked 1,133 low-pressure centers, assigned them to one of nine areas of origin, and mapped average trajectories for each month. "If all roads lead to Rome," he concluded, "all storms and weather tracks lead to New England."[100] Although high-pressure cells could impede, accelerate, or divert low-pressure cells from a smooth trajectory, the ultimate destination was clear.

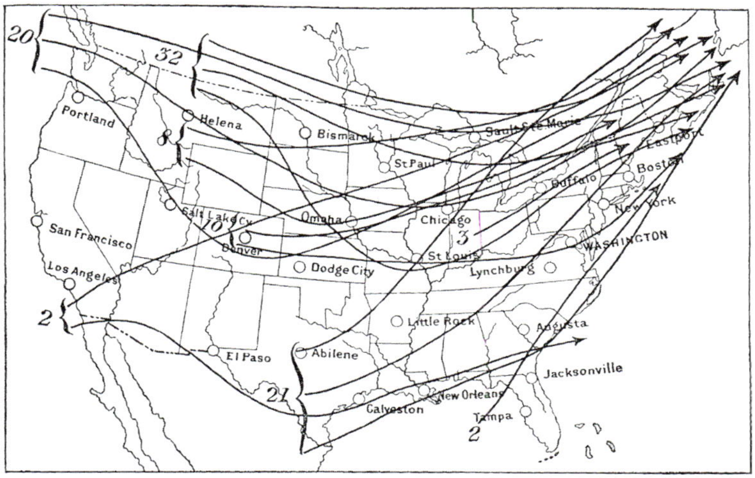

Figure 5.11. Numbers on Frank Bigelow's map of typical storm tracks for January show the total number of storms originating in each region between 1884 and 1893.

Operational meteorology has changed radically since the days of Abbe and Bigelow, with differences particularly prominent in the type and precision of instruments, the number of vantage points and frequency of observations, the complexity of the underlying theory, and the speed with which data flow through the network and onto maps. Even so, rampant technological progress has not altered the underlying strategy of knowledge production grounded in a network of synchronous observations. What's different is the emergence of a dynamic network that not only integrates data from Doppler radar, geostationary satellites, and automatic measuring devices but also uses computer models to predict the state of the atmosphere days or weeks ahead.

Most prominent among these many advances is Doppler radar, which made "nowcasting" an indispensable adjunct to forecasting. Doppler radar is based on Christian Doppler's (1803–53) discovery that sound from a source approaching the observer has a shorter wavelength than sound from a source moving away. Think of the change in pitch of the air horn as a rapidly moving train moves past. Conventional weather radar merely determines distance from the radar tower to particles of precipitable moisture—radar signals can be tweaked to bounce off hailstones and water droplets, and distance is calculated from the lapsed time between transmission and return. By contrast, Doppler radar can detect rotation in a tornado because particles on one side of the narrow storm's center are moving rapidly toward the radar tower while those opposite are moving away.

For more than a dozen years, the Next Generation Weather Radar (NEXRAD) map has been my favorite map. Nowadays, anyone with a radar app on their mobile device can watch an approaching line of severe storms, with high winds and hailstones. And when tipped off by algorithms that vigilantly scan the radar map for deadly rotating winds, meteorologists can monitor the movement of tornadoes, detected well before someone spots a funnel cloud and phones in a report. Everything is networked: the tower-mounted radar instruments that monitor precipitable moisture and wind speed and contribute to national and regional radar maps, the weather forecast offices responsible for issuing alerts to the media and over the internet, and All Hazards Radio, which broadcasts a silent signal that activates customized radio receivers within geocoded danger zones.[101] A receiver that detects a signal matching its location turns on the audio and sounds a tone alarm to get residents' attention.

Among the more dramatic impacts of electronic weather surveillance was the progressive revamping of forecasting operations. By the late 1920s, the teletype had replaced the telegraph, and in 1940 Congress recognized the increased importance of aviation by moving the Weather Bureau from the Department of Agriculture to the Department of Commerce.[102] In the late 1950s, computer models were promising more reliable predictions. By 1967, forecasting operations had evolved to include separate centers focused on hurricanes and tornadoes, and a three-tier nested hierarchy of twenty-two offices prepared forecasts looking ahead forty-eight hours and covering one or more states. Fifty-nine offices refined these predictions for smaller regions, and 235 offices issued forecasts tailored to local areas.[103] Within the conterminous United States, the 232 sites with a local forecast office (some with higher-level responsibilities) roughly approximated the pattern of urban centers (figure 5.12, top).

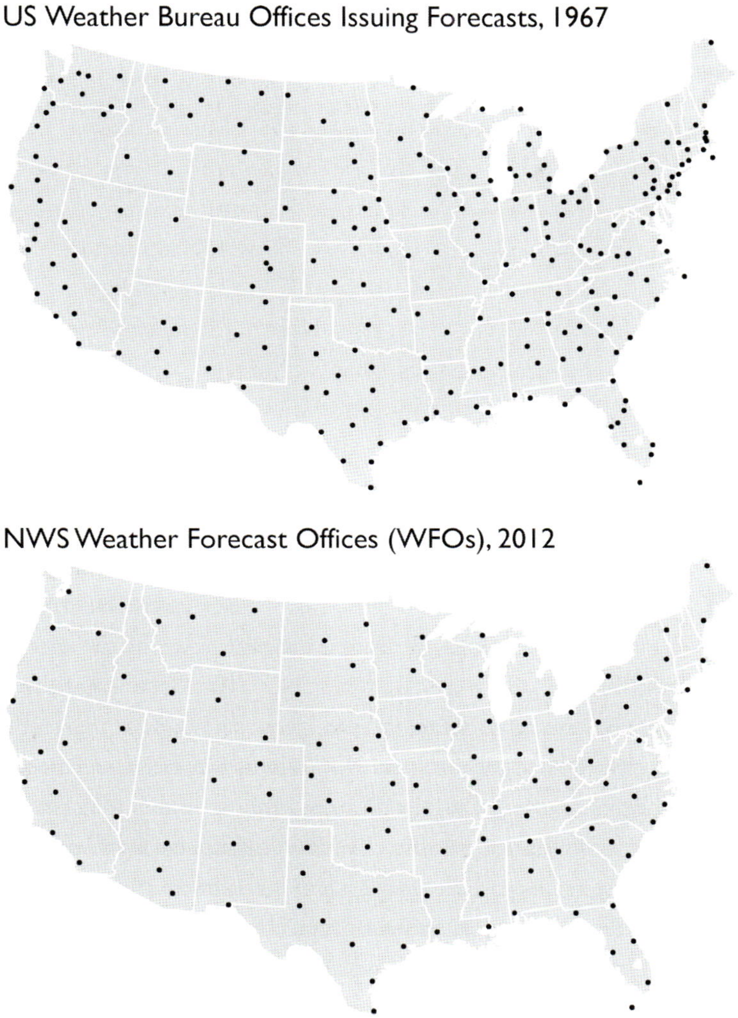

Figure 5.12. Weather offices in the conterminous United States issuing forecasts in 1967 (*top*) and 2012 (*bottom*).

Acronyms proliferated. In 1970, the Weather Bureau became the National Weather Service (NWS) when Congress moved geodetic surveying and weather forecasting into the newly created National Oceanic and Atmospheric Administration (NOAA). By 1989, changing technology had led to fifty-two Weather Service Forecast Offices (WSFOs) run by professional meteorologists who oversaw 204 Weather Service Offices (WSOs) staffed mostly by technicians and observers. That year became significant when NOAA announced the Modernization and Associated Restructuring program, which replaced this two-tiered system with 122 Weather Forecast Offices (WFOs), all staffed with professional meteorologists and all providing 24/7 coverage.[104] Cities such as Rochester and Syracuse lost their local offices to a network of more evenly spaced WFOs (figure 5.12, lower), each with its own WSR-88D radar instrument.[105]

WSR-88D is shorthand for Weather Service Radar-1988 Doppler, which was markedly more powerful in detecting precipitable moisture than earlier radars.[106] Its better-known acronym, NEXRAD, describes a network configured to minimize gaps in nationwide coverage. The 143 stations in the conterminous United States contribute to the composite radar images prominent on cable television and mobile weather apps.[107]

NEXRAD's ability to detect weather targets depends on a storm's altitude, intensity, and size. In general, detection is more reliable for tall, wide storms with large, dense precipitation particles. Typical median ranges are 386 km (240 mi.) for a hurricane eyewall, 230 km (143 mi.) for a supercell mesocyclone, and 145 km (90 mi.) for lake-effect snow.[108]

NEXRAD's improved coverage was evident in Washington and Oregon (figure 5.13), where the number of radars increased from two to six, and the area not covered shrank dramatically. Circles centered on these radar sites have an admittedly arbitrary radius of 230 km beyond which some (but not all) phenomena fall below the lower edge of the radar beam: as distance from the radar tower increases, the upwardly tilted beam rises above the ground and Earth curvature adds to the vertical separation. The irregular boundaries of coverage areas reflect mountainous terrain that interferes with detection in remote, sparsely populated areas. NEXRAD deployment reflects a compromise between comprehensive vigilance and prudent spending.

Restructuring involved multiple principles exemplified in figure 5.14, which focuses on upstate New York. Note that all the WFOs in the left panel share the name of a radar site in the right panel. Although WFOs and their radars were usually not next-door neighbors, they were often sufficiently close to be considered collocated. Binghamton, New York's, radar tower is less than a mile from the forecast office, whereas Albany's radar is 14 mi. away, on a locally prominent

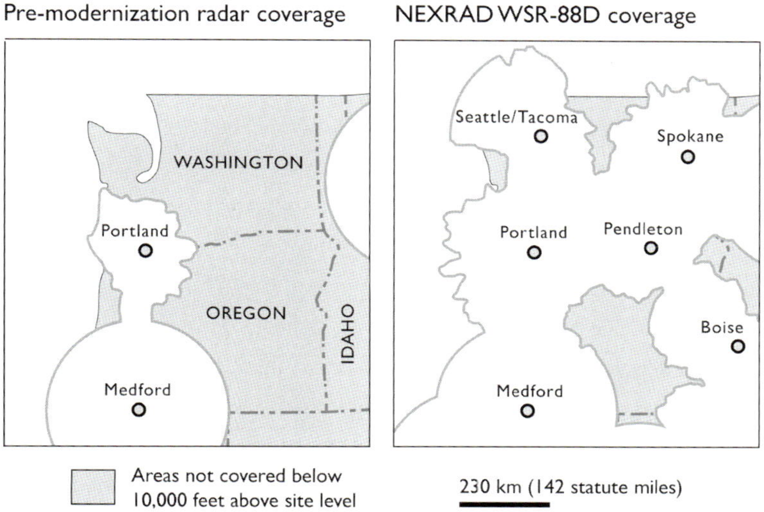

Figure 5.13. Weather radar sites and generalized coverage areas for Washington and Oregon before and after the NWS Modernization and Associated Restructuring program.

ridge.[109] Note, too, the additional radar at Montague, New York, which is owned by the Defense Department and maintained by Army personnel at Fort Drum, 50 mi. to the north but mostly controlled by the WFO at Burlington, Vermont. In 1998, the instrument was moved to Montague from Griffiss Air Force Base, near Rome, to provide better coverage of lake-effect snow off Lake Ontario, to the west, as well as better protection for Fort Drum.[110] The NEXRAD network in the conterminous United States includes twenty-one military radars and 122 owned by the NWS.[111]

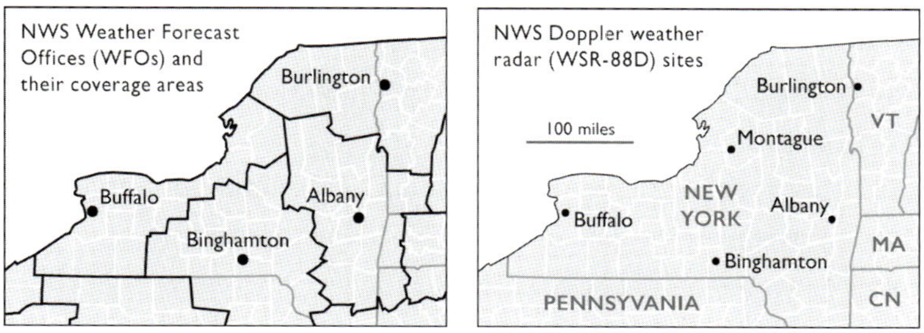

Figure 5.14. WFOs, forecast coverage areas, and Doppler radar locations within and around upstate New York.

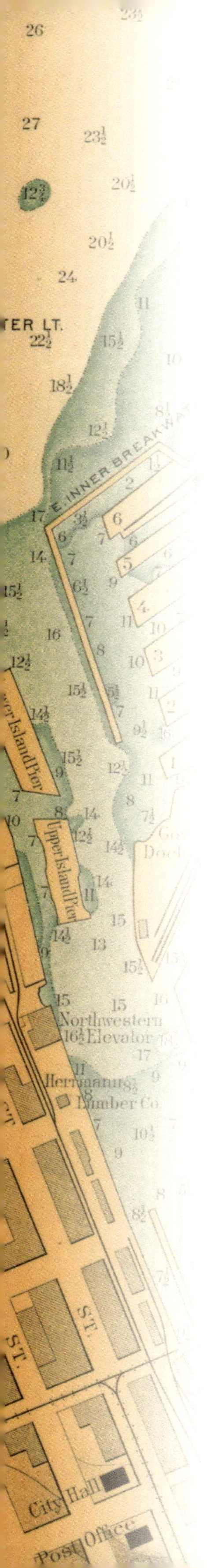

Note, too, that each WFO has a forecast area composed mostly of whole counties and that forecast areas often cross state boundaries.[112] As the Buffalo WFO demonstrates, proximity was not the sole factor in assigning counties: Buffalo's forecast area reaches eastward to include all counties in the Lake Ontario snow belt, which is vulnerable to heavy lake-effect snow from midfall through early spring. Buffalo meteorologists specialize in forecasting the lake effect and advising local officials of its likely consequences.[113]

Another transformative aspect of NWS modernization is the Automated Surface Observing System (ASOS) network.[114] As its name implies, ASOS's computer-controlled sensors record temperature, barometric pressure, wind speed and direction, dew point (an indicator of humidity), wind gusts, and other weather measurements relevant to a site. In addition to sharing results with the WFO network in near real time, each ASOS unit has a data-to-speech apparatus that broadcasts observations to pilots and provides dial-in telephone reports. Because aeronautics depends on timely weather information, ASOS units are commonly located at airports—in and around upstate New York, all cities with scheduled airline service have an ASOS site (figure 5.15).[115] Collectively, the nation's more than nine hundred ASOS sites constitute an electronic observation network notably denser than any prior configuration of stations staffed by human observers.[116] Even so, the NWS also receives data relevant to climate science from a network of over ten thousand volunteer observers.[117]

In sum, operational meteorology is now the embodiment of the map/network nexus. As Brandes demonstrated in the early nineteenth century, mapping synchronous atmospheric observations acquired at dispersed locations is a prerequisite to understanding weather. Once the electric telegraph fostered timely maps offering insights about the next day's weather, networks became important in disseminating results. When nowcasting demonstrated the value of a cost-effective network of reliable high-resolution Doppler radars, the network of forecast offices was reconfigured around this remarkably incisive instrument, and the network of measurements and data became the network of interpretation and public understanding.

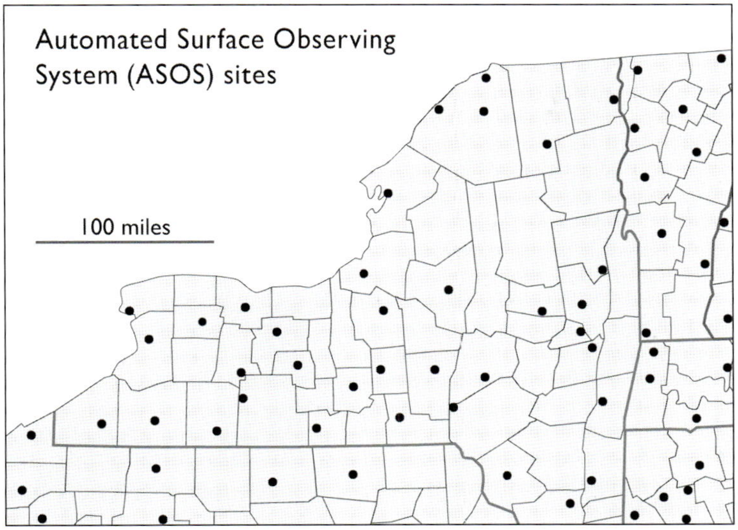

Figure 5.15. ASOS locations within and around upstate New York.

6

TOPOLOGY

Here's what topology is about: draw a map on a sheet of rubber and stretch it various ways without tearing the sheet, and however much you've altered sizes and shapes, counties that abut still abut, roads that intersect still intersect, and a town with seven lakes still has seven lakes, even though you might no longer recognize their shapes. Herein lies the fundamental distinction between *geometry*, which is concerned with distances, angles, and area, and *topology*, which focuses on the connections and adjacencies that help computers solve routine mapping problems, such as finding an efficient route to your vacation rental or counting residents in your town. Many important applications that make maps depend on data reflect topological relationships. This chapter's simplified but unavoidably technical examples are worth a close read if you want to understand the diverse ways in which network data make maps legible to computers and the ways improved computing has altered the organization and use of data. My focus on basic conceptual strategies for structuring data is not intended to reflect the current (and ever changing) sophistication of cartographic software but to clarify the strategies behind networks and cartography.

To appreciate the need for topologically structured data, you must first understand how and why computers treat maps as lists of coordinates and linkages. If you recall working with graph paper in middle school, you surely understand how lists of coordinates promote the storage, projection, and display of coastlines, boundaries, and other linear features—think of a connect-the-dots puzzle with an x,y list instead of numbered dots. Less obvious, especially if you've never programmed a computer, is the use of lists to describe areas that touch or roads that join. These lists are the basis of topologically structured data, and even if you've never seen one, these lists are easy to appreciate once you acknowledge that computers must be tricked into seeing what's readily obvious to the human eye.

Consider figure 6.1, which uses the boundary between Kansas and Missouri to illustrate how lists portray topology.[1] Thick lines in the upper half of the image highlight this shared boundary as part of each state's perimeter, represented electronically as two separate lists of coordinates. Although present-day approaches to displaying maps electronically are more sophisticated, this simplified example provides a straightforward (and historically relevant) introduction to general principles. In the early days of what was called *computer cartography*, a technician would mount a map on a digitizing tablet and track around each state, recording coordinate pairs in a clockwise sequence. Capturing the Kansas–Missouri boundary twice—once for the perimeter of Kansas, and again for the perimeter of Missouri—was inefficient and often aesthetically flawed: it doubled the amount of memory required, and the sinuous line captured for Kansas was seldom an exact replica of the line captured for Missouri. A close look would usually reveal overlaps or gaps when the common boundary was plotted twice, once for each state. This method left two choices: accept the blemishes of quick-and-dirty digitizing or clean the data with a tedious manual edit.

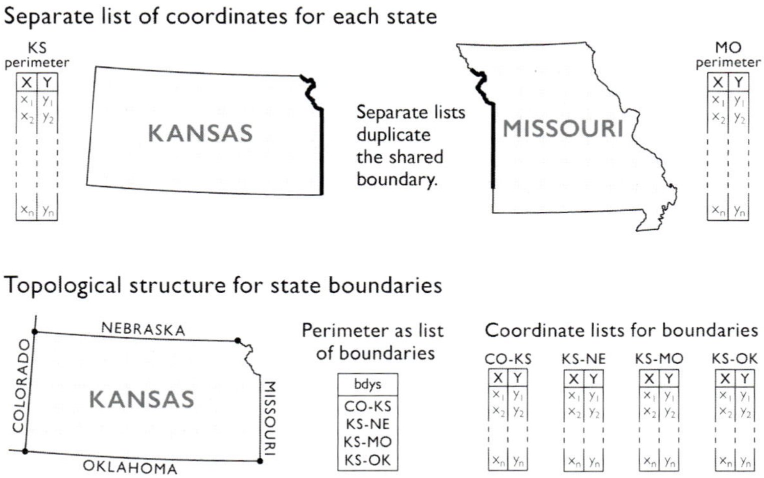

Figure 6.1. Example of topological principles applied to the Kansas–Missouri border.

Tedious for a human but just right for a computer. Efficient software can make the pieces match and yield the topologically structured data described in the lower half of the diagram, which focuses on Kansas.[2] Note that a single list describes each shared boundary, and a sequenced list of shared boundaries describes each perimeter. Also note that these shared boundaries connect at nodes (represented by black dots), which ensures that the endpoints of every boundary segment have identical coordinates.

Topological data have their own jargon. Nodes are sometimes called *vertices*, the boundary segments that connect at nodes are *arcs*, and an organized collection of arcs and their coordinates is *vector data*, derived from the mathematical definition of *vector* as a straight line headed in a specific direction between two points. Although "arc" usually refers to a curved feature represented by a list of points, an arc is sometimes just a straight line. The term should not be confused with the *arc of a circle* (a part of the circumference) or with the *arcs of triangulation* discussed in chapter 1—comparatively recent fields such as GIS (*geographic information science* in some contexts, *geographic information system* in others) create confusion by appropriating terms with established meanings in other fields. Less problematic is *polygon*, which refers to a closed two-dimensional feature represented by a list of coordinates that describes a connected sequence of straight-line sides, vectors, or *edges*. Unlike the simple polygons of middle school geometry, a data polygon can have thousands of sides. Although polygons with highly irregular boundaries posed a computational challenge in the early 1960s, improved computers could rapidly and reliably calculate any polygon's center, area, and perimeter. Nowadays, even closed curves drawn by hand are called "polygons."

More challenging was the common occurrence of *enclaves* or *outliers* on environmental maps, illustrated by the excerpts reproduced in figure 6.2 at their original scale. The soils map on the left is from a US Department of Agriculture report prepared by soil scientists who spent years in the field delineating areas with a similar soil, labeled on the map with a two- or three-letter code.[3] Several small polygons such as the patch of Croghan loamy fine sand (CrB) near the left edge lie wholly within a larger polygon, in this case a band of Naumburg loamy fine sand (Na). And just above the center, an enclave of Ontario Loam (OgB) lies inside an enclave of Hilton loam (HlB), which in turn is surrounded by a more extensive occurrence of Appleton loam (AoA). In GIS parlance, these enclaves are *islands*, and a group of adjoining polygons surrounded by an outer polygon is a *compound island*.[4]

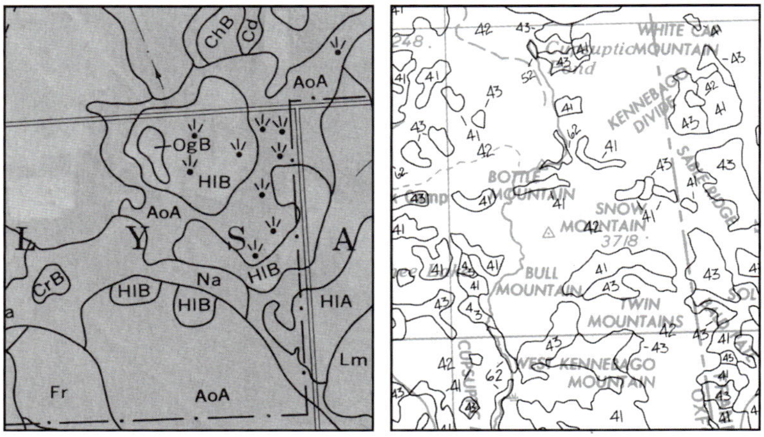

Figure 6.2. Excerpts from a 1:20,000 soils map (*left*) and a 1:250,000 land-use/land-cover map (*right*) contain examples of simple and compound islands.

Islands are more common in the excerpt on the right, from an area in western Maine near the Quebec border, as described on a USGS land-use and land-cover map. Two-digit attribute codes identify the map's meticulously labeled polygons as deciduous (41), evergreen (42), or mixed-forest (43) land, and in a few cases as a lake (52) or a nonforested wetland (62). Evergreen forest is the predominant land cover but islands of other categories abound. Toward the upper right is a compound island within which adjoining patches of deciduous (41) and mixed (43) forest enclose a small patch of evergreen forest (42). The map would be much more fragmented if USGS rules for land-cover mapping did not set 16 hectares (40 acres) as the minimum mapping unit for forested land.[5]

Because environmental and land management studies use electronic maps to tally the total land area for individual categories, polygons containing islands require special treatment. Islands are holes, after all, and their individual areas must be deducted from the area calculated for the bounding polygon's perimeter. A straightforward strategy for coping with complex configurations assigns a unique index number to every polygon in the dataset, including islands and compound islands, and maintains a separate list of included islands for every bounding polygon. This strategy readily accommodates islands within islands because each outer island has its own inclusions list. For the nested islands in the soils map (figure 6.2, left), the area of the inner OgB island is deducted from the area within the perimeter for the outer HlB island, deducted in turn from the area of the surrounding AoA polygon. As part of the database's topological structure, lists of inclusions ensure reliable tallies.[6]

Plotting land polygons on a map can be a conceptually uncomplicated process when inclusion lists sort polygons into layers so that an island can overwrite part of its surrounding polygon. An electronic version of the map is composed as a high-resolution image before rendering it on a laser printer or display monitor.[7] Sequence is important: the software begins with a blank artboard, to which it adds the interiors of bottom-layer polygons, such as AoA on the soils map (figure 6.2, left). It then adds a second layer of islands, such as HlB, and maybe a third layer, such as OgB, if these islands contain their own islands, and so forth. The procedure then adds all boundary lines followed by any text or frame-of-reference features, such as political boundaries, important highways and railroads, and place-names.

Producing a high-resolution map often requires converting vector data (lists of coordinates representing arcs and nodes) to raster data (akin to the pixels displayed on a printer or monitor). Unlike topologically structured vector data, raster data consist of cells or pixels arranged in a grid of numbered rows and columns. Algorithms for converting vector data to raster data, and vice versa, are important because each format has distinct advantages. Although the early 1980s witnessed the often heated debate over which data model is better, dramatic advances in processing speed and electronic memory as well as hybrid vector-raster data have made this differentiation a nonissue.[8] That said, the network topology of vector data is the dominant approach to producing geographic data, and the grid framework of raster data is dominant in publishing maps and integrating satellite and other overhead imagery.

Derived from *rastum*, Latin for *rake*, *raster* is the natural format for image data, captured line after parallel line by a desktop scanner or satellite sensor. Image-processing software is used to manipulate and interpret satellite data, and when multiple coverages representing land use, vegetation, or soil can be registered to the same grid, a raster format is efficient for analyzing land-use change or monitoring a contagious infestation affecting trees or crops. If a formula allows the quick conversion of latitude and longitude or plane coordinates to row and column numbers, raster data are well suited to queries about conditions at specific points, and if the grid cells are suitably small, a raster format is useful for elevation and landownership data. Although a grid allows for the ready analysis of relationships among neighboring cells (and even among neighbors a few steps removed), a raster format is awkward for representing streets, railways, and other phenomena more logically treated as lists of points.

Vector data are useful when an insurance company or city planner needs to know whether a structure with known coordinates is in a flood zone, such as polygon A in figure 6.3. Reliable point-in-polygon algorithms make this determination straightforward: simply test each flood polygon to see whether it contains the structure.[9] Because this so-called "brute force" approach can be computationally cumbersome when the database consists of tens of thousands of polygons, some with hundreds of sides, a more efficient strategy uses *minimum bounding rectangles* (figure 6.3). This method screens for polygons sufficiently close to warrant a full point-in-polygon test: if the structure is outside a polygon's "bounding box," merely move on to the next polygon.[10] Simple geography offers further efficiency when polygon data are organized by region, state, county, and town. For example, if the polygon in question is somewhere in Florida, starting in the Pacific Northwest and working eastward and southward is pointless.

Geographic screening also expedites use of the massive National Elevation Dataset (NED).[11] When the nationwide 1 m digital elevation model (DEM) is complete, the Geological Survey's online Elevation Point Query Service will return an elevation accurate to within about 10 cm for any location identified by latitude and longitude.[12] The whole-number portions of a point's decimal latitude and longitude narrow the search to a specific part of a large database organized by 1° × 1° tiles—think of the square marble or ceramic tiles on a bathroom floor but with the curved sides of a sphere. These tiles are subdivided, in turn, into square DEM blocks that are 10 km (6.21 mi.) on a side. Elevations are recorded for the intersections of north–south and east–west grid lines just 1 m apart—for a 10,000 × 10,000 grid that's 100 million elevations in a single block.[13] The 1 m DEM is an enormous improvement in both horizontal and vertical accuracy over elevations estimated by interpolating from contour lines on a printed topographic

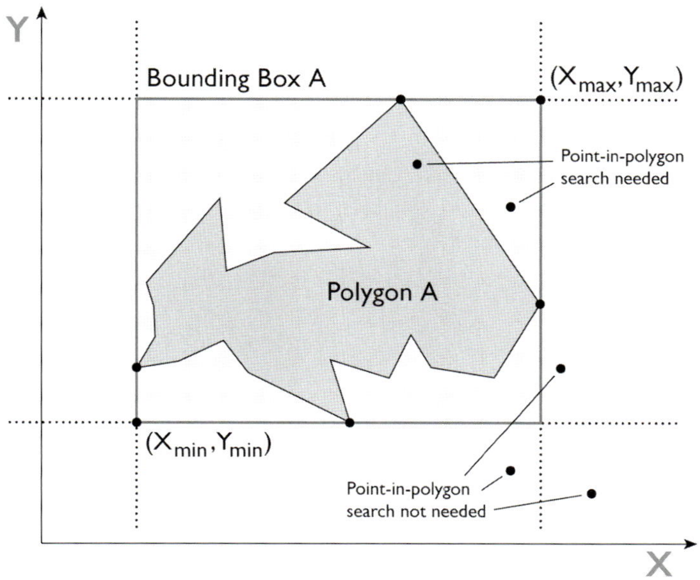

Figure 6.3. Minimum bounding rectangles expedite searching by screening for polygons with a bounding box that contains the point in question.

quadrangle map.[14] For example, if the contour interval is 10 ft., map accuracy standards offer only a 90 percent guarantee that a map-based elevation estimate is within 5 ft. (152 cm) of the true value.[15]

Although the raster-mode DEM is the predominant format for high-resolution elevation data, the aptly named *triangulated irregular network* (TIN), a topologically structured vector model of terrain, can create revealing maps with far fewer points.[16] The TIN model treats a three-dimensional elevation surface as a lattice of triangles joined at representative points along a ridge or valley line or along noteworthy breaks in slope. Like the topologically structured arc-polygon example in figure 6.1, the TIN model relies on lists to represent topological structure. In figure 6.4, a single consolidated list describes relationships among a hypothetical network's triangles (*A* through *P*), nodes (1 through 13), and edges (sides of triangles, not labeled). Each row represents a triangle, the Nodes column lists the three nodes that define the triangle, and the Neighbors column lists the triangles that abut along the three edges. An asterisk (*) in this last column indicates that one edge is on the network's perimeter. To clarify the strategy, a diagram at the

bottom highlights the relationships for triangle *L*, defined by nodes 6, 12, and 13 and abutting neighboring triangles *K*, *M*, and *P*. An additional list (not shown) contains x,y,z coordinates for the nodes.

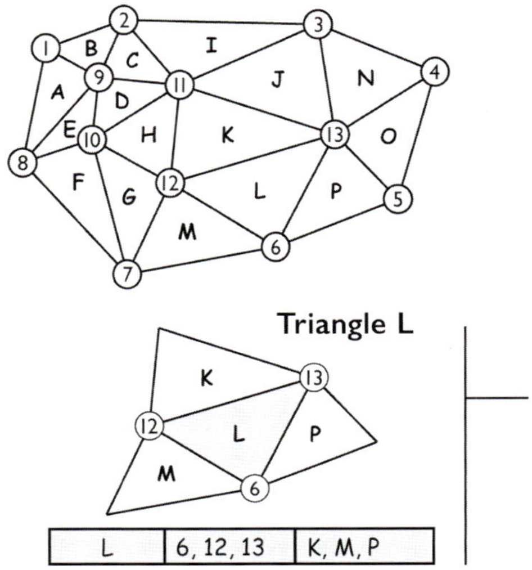

Topological structure

Triangle	Nodes	Neighbors
A	1, 8, 9	B, E, *
B	1, 2, 9	A, C, *
C	2, 9, 11	B, D, I
D	9, 10, 11	C, E, H
E	8, 9, 10	A, D, F
F	7, 8, 10	E, G, *
G	7, 10, 12	F, H, M
H	10, 11, 12	D, G, K
I	2, 3, 11	C, J, *
J	3, 11, 13	I, K, N
K	11, 12, 13	H, J, L
L	6, 12, 13	K, M, P
M	6, 7, 12	G, L, *
N	3, 4, 13	J, O, *
O	4, 5, 13	N, P, *
P	5, 6, 13	L, O, *

Figure 6.4. A TIN represented by its map (*left*) and a table describing topological relationships for its triangles (*right*). The shift in typeface from Gill sans serif to Comic sans serif for figures 6.4, 6.5, and 6.11 reflects the need for a clearer rendering of these illustrations' abundant numerals.

The list in figure 6.4 can be supplemented by a second topological list focusing on the edges. As shown in the left half of figure 6.5, the table has a separate row for each edge, identified concisely by the pair of triangles it separates. For example, the three sides of triangle *L* are listed in three separate rows as *L-K*, *L-M*, and *L-P*. Another column indicates the pair of nodes each edge connects: in this example, 12-13, 6-12, and 6-13, respectively.

Synchronized inspection of the triangle- and edge-based tables can guide the threading of contour lines through the network, which was an early use of the TIN data model. Contour threading follows the simple rule in which a contour that enters a triangle must leave it along one of the other two sides. The rule refers to surface elevation at the nodes, retrieved from the x,y,z list of coordinates shown next to each node in the right half of figure 6.5. Once a contour enters a triangle, the elevations determine the edge through which it must exit.[17] The rule is

Edge/Nodes relations

Edge	Nodes	Edge	Nodes
A-*	1, 8	I-J	3, 11
B-*	1, 2	J-K	11, 13
I-*	2, 3	H-K	11, 12
N-*	3, 4	D-H	10, 11
O-*	4, 5	E-F	8, 10
P-*	5, 6	F-G	7, 10
M-*	6, 7	G-H	10, 12
F-*	7, 8	G-M	7, 12
A-B	1, 9	L-M	6, 12
B-C	2, 9	K-L	12, 13
C-D	9, 11	J-N	3, 13
D-E	9, 10	N-O	4, 13
A-E	8, 9	O-P	5, 13
C-I	2, 11	L-P	6, 13

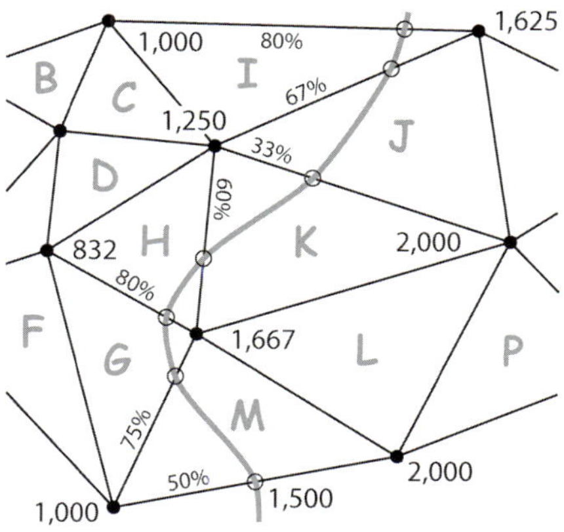

Figure 6.5. Table (*on the left*) describes topological relationships for the edges of the triangulated network on its right, which is part of the network in figure 6.4. Diagram (*on the right)* illustrates the threading through the network of a contour line with an elevation of 1,500. Numbers at the nodes are *elevations*, and numbers along selected edges indicate the contour crossing's percentage distance away from the indicated node.

straightforward: a contour will not cross an edge unless its elevation falls between the elevations of the two bracketing nodes. Once the exiting edge is determined, an interpolation rule specifies the crossing point.[18] For example, if the contour elevation is halfway between the elevations of the departing edge's nodes, the crossing will occur at the edge's midpoint. Moreover, the smaller the vertical difference between a contour and a node, the smaller the horizontal difference between the crossing and the node. To avoid unnatural abrupt bends where it crosses an edge, I had to smooth the gray contour line for it to look more realistic.[19]

An arguably more important application of TIN data is the rapid interactive construction of oblique three-dimensional terrain maps, which depends on an efficient, reliable, on-the-fly strategy for removing hidden lines. Although each triangular facet of the elevation surface can in principle be projected onto a two-dimensional computer screen, some facets must be ignored altogether because they lie outside the rectangular viewing frame, some facets must be suppressed because they are hidden behind other facets closer to the viewer, and facets that are only partly visible must be overwritten by facets closer to the viewer.[20] Similar principles could drive an interactive display based on a high-resolution DEM, but the much larger number of data points might make this computationally

cumbersome process too slow for an interactive display in which the scene is constantly changing. TIN accelerates real-time three-dimensional terrain mapping by requiring far fewer data points than an equivalent DEM representation.

Figure 6.6 attests to this computational efficiency with two different views of the same terrain database created seconds apart by an online map server. Although the junctions of triangles are not apparent because of smoothing and the moderately high spatial resolution of the data, the facets have been shaded to simulate differences in the amount of light reflected from a light source on the upper left. In addition to labels and symbols representing vegetation cover, streams, roads, boundaries, and other features, the map is further enhanced by contours threaded across triangular facets facing the viewer. As this example suggests, contour lines on a 3D map can either enhance or replace hillshading.

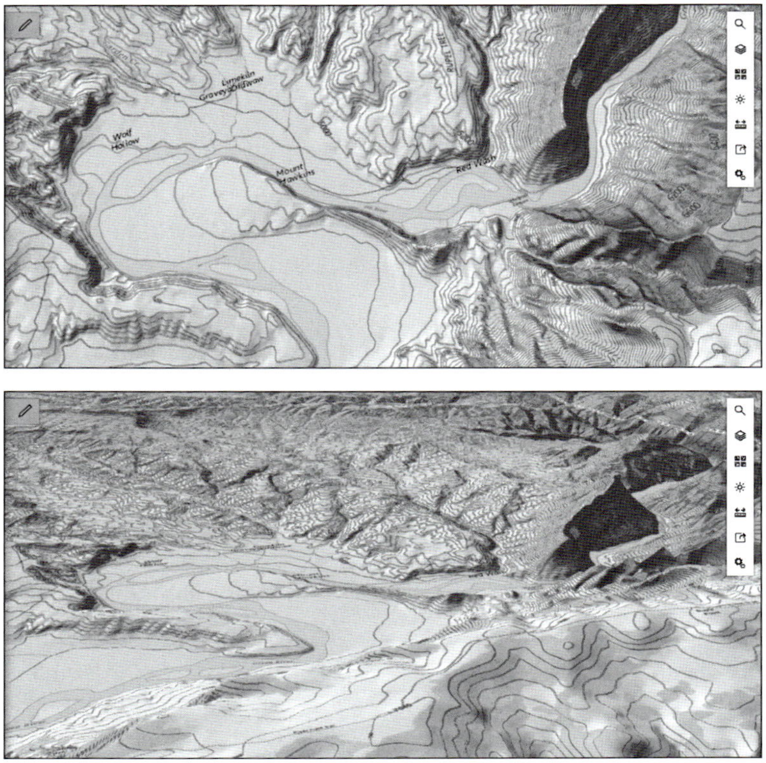

Figure 6.6. High-angle (*top*) and low-angle (*bottom*) oblique views of terrain in part of Dinosaur National Monument, created interactively with ArcGIS® Online.

Topological principles for threading contours through triangulated networks are equally useful in discerning optimum routes across electronic street maps. Instead of focusing on triangles joined at nodes and edges, topological treatments of the urban street grid focus on city blocks and the street segments that separate them. In network terms, street intersections are *nodes*, street segments between successive nodes are *arcs*, and city blocks are *polygons*. Similar conceptually to the arc-polygon framework used for land polygons and state boundaries (see figure 6.1), the street grid scheme has been extended beyond city limits to include rural roads and expanded to encompass streams, shorelines, railroads, and political boundaries. Arcs in the city and its suburbs are enhanced with street names and addresses, and blocks are assigned unique numbers by the US Census Bureau, which uses them for its decennial headcount. Blocks are the fundamental building blocks of congressional redistricting—the reason why we have a census—and block data aggregated by census tract are useful for mapping and statistical analysis.[21] Moreover, an appreciation of topology's impact on census mapping is fundamental to understanding the rampant partisan gerrymandering examined in the following chapter.

Although topologically structured road maps nowadays serve travel websites and GPS navigation systems, the Census Bureau pioneered the concept primarily to streamline the 1970 Census, which introduced the mail-out/mail-back questionnaire, sent to 60 percent of the nation's households.[22] Automation was not new—the bureau had used punched cards for tabulating the 1890 Census—but census taking had remained a largely manual process requiring an army of door-to-door census takers who recorded householders' responses in cursive script that continues to challenge optical character recognition (OCR) software. The Third Industrial Revolution, epitomized by post–World War II advances in digital computing, brought about the high-speed printers that addressed the questionnaires and the optical scanners and software that tabulated fill-in-the-box responses.[23] Although electronic data processing was helpful, the run-up to a nationwide enumeration was a massive clerical chore that included numbering census blocks, revising tract boundaries, and field-checking the data. Preparations encompassed acquiring and comparing address lists from direct-marketing firms and field-checking a master address list to make certain all households were

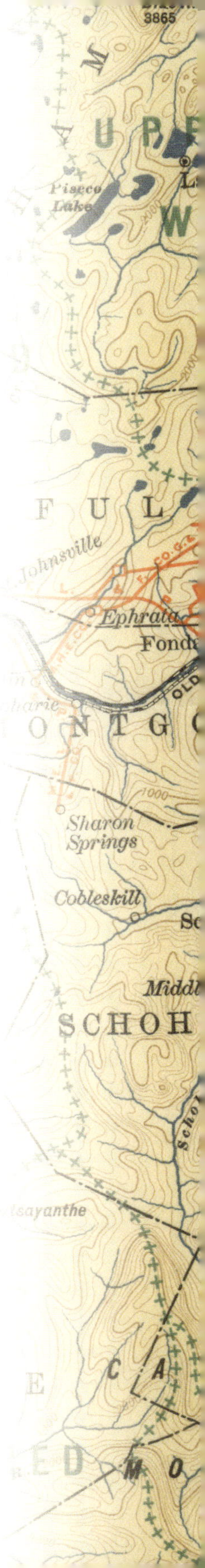

included. An indispensable product of this effort was the address coding guide, or ACG, used to relate an address to a specific block and census tract.

By itself, the ACG is not a topologically structured map. As described in the right half of figure 6.7, an alphabetical list of streets is subdivided by block, with separate records for even- and odd-number addresses—an essential split because residences on opposite sides of the street belong to different blocks. In addition to the block number, records for each block face, one side of a street segment, include the low- and high-number addresses along with the tract number, five-digit zip code, and other numbers such as the ward and congressional district, used for special tabulations. To match a household with the correct block, a computer merely jumps to the part of the ACG with the street name, jumps again if necessary to distinguish between a street, avenue, drive, road, or parkway sharing the same name, and finds the address range bracketing the address in question. For example, 308 Fleet Street, on the even-number side of the street between 300 and 328, would be in block 82 and tract 12.

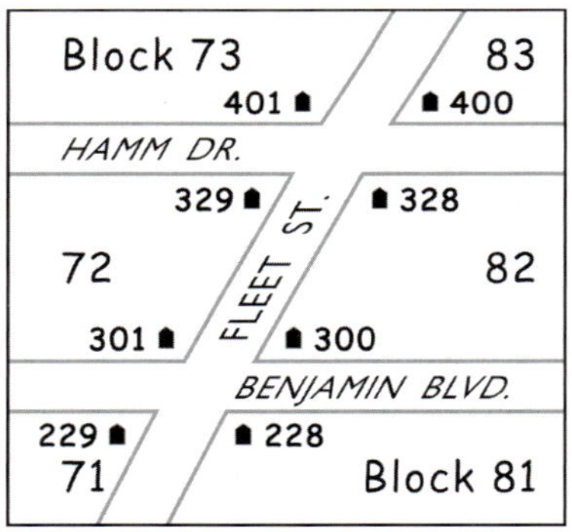

Street		Address		Census	
Name	Type	Low	High	Block	Tract
Fleet	St.	200	228	81	12
Fleet	St.	201	229	71	12
Fleet	St.	300	328	82	12
Fleet	St.	301	329	72	12
Fleet	St.	400	428	83	12
Fleet	St.	401	429	73	12

Figure 6.7. Hypothetical street map (*left*) and corresponding portion of the ACG (*right*).

Improved data formats with wordy names made acronyms unavoidable.[24] For the 1970 Census, an ACG became a GBF (geographic base file) when the Census Bureau added x,y coordinates for the nodes and a set of topological features known as DIME, for dual independent map encoding. DIME enhancement introduced line features such as railroads and streams and whatever additional points were needed for a slightly more realistic treatment of cul-de-sacs and curved streets. As illustrated by the hypothetical example of Victory Circle (figure 6.8, top) in a 1976 supervisor's manual, "too many nodes" were considered "excessive and unnecessary," whereas "a reasonable representation" of a street with a noticeable kink or bend required an additional node (number 8 in figure 6.8, bottom).[25] Increased curvature made the 1980 data less graphically clumsy than their bare-bones 1970 counterpart, but GBF/DIME coverage was largely confined to metropolitan areas. The 1990 enumeration brought markedly expanded coverage after the Census Bureau collaborated with the Geological Survey on a nationwide digital map called TIGER, for topologically integrated geographic encoding and referencing. Comparable in detail to printed USGS 1:100,000 topographic maps, digital TIGER/Line® files covered the entire middle forty-eight states, also known as CONUS, for conterminous [contiguous] United States.[26]

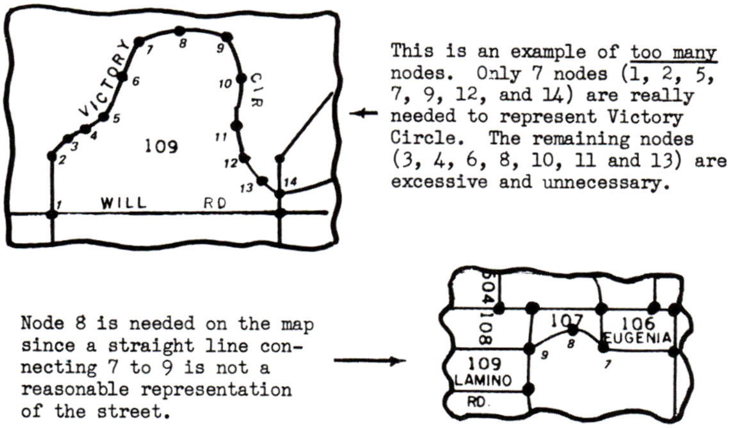

Figure 6.8. Graphic examples of too many (*top*) and too few (*bottom*) nodes, from a 1976 Census Bureau supervisor's manual for creating GBF/DIME data.

The prime reason for adding DIME features to an ACG was an automated strategy for confirming the correctness of an electronic street map. Devised by James Corbett (1909–98), a mathematician at the Census Bureau, a set of algorithms ferreted out inconsistencies or errors that arose when clerks numbered the map's nodes, street segments, and blocks; entered these numbers onto coding forms; keypunched the information onto cards that were read onto magnetic tape; and edited the data to correct errors or accommodate changes to the street network or tract boundaries—mistakes could arise at any stage.[27]

Corbett's application of topological principles treated each street segment as not only bounded by the nodes it connected but also cobounded by the blocks it separated—hence, the "dual independent" in DIME.[28] He also assigned each street segment a direction and coded its bound/cobound relationship according to the following quadruple:

Node 1, Node 2 : Block *A*, Block *B*,

indicating that in the direction from Node 1 to Node 2, Block *A* lies to the right and Block *B* lies to the left. This relationship holds when the direction is reversed and entities on each side of the colon are swapped, as in

Node 2, Node 1 : Block *B*, Block *A*,

which indicates that in the direction from Node 2 to Node 1, Block *B* is on the right and Block *A* is on the left.

Coding individual street segments is usually straightforward. Because DIME records are similar to the paired rows in an ACG, they include the street's name and type, the numbers of adjoining blocks and tracts, and low- and high-address numbers for opposite sides of the street. The progression of addresses provides the segment's direction, which in turn distinguishes right from left. For the 300 block of Fleet Street in figure 6.7, block 82 is on the right and block 72 is on the left.

Not all streets fit the conventional assumption about how street addresses work. A manual for coding DIME files recognized anomalies such as the offset streets and inconsistent numbering in figure 6.9. In a strictly rectangular street grid, Bay Street would run directly into Oak Street (perhaps without a change in name), Elm Street would align with Sea Street, and the three-block stretch of Washington Street between Olive Street and Pine Street would require only four nodes (rather than six) and only six block faces (rather than ten). Note the added irregularity of even and odd numbers on the same side of the street: a rare but presumably plausible situation anticipated by census officials who developed guidelines that suggest:

> It may be necessary to divide the address numbers for each side of a segment into actual address ranges (both even and odd) which do not overlap any other address numbers on the other side of the segment or in any other segment.[29]

For the segment between nodes 43 and 44, this guidance calls for three address ranges on the left: 37 (because it's an odd number on the even side of the street), 38–44, and 48 (because it's within the range of odd-number addresses on the other side of the street). What seems like obsessive hair-splitting reflects the need to assign addresses to the correct block. In this example, because number 46 is not within any of the address ranges for the left (even) side of the street, a search would then look on the right (odd) side and find it in block 410, opposite the 38–44 range for block 305. Additional guidance recognized the possibility of address numbers out of sequence.[30] Computer algorithms require this degree of specificity.

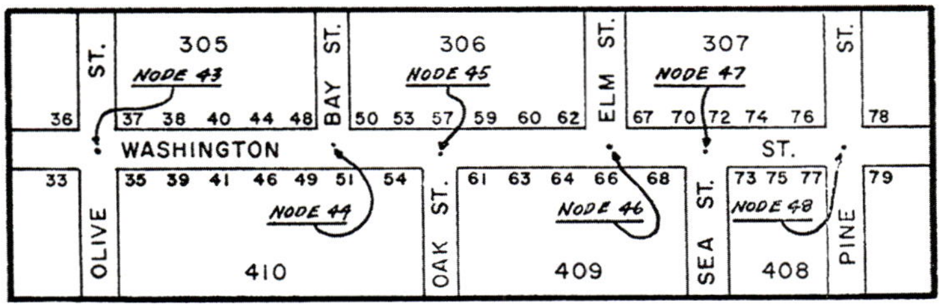

Code (e)	From node: Map No. (f)	From node: Node No. (g)	To node: Map No. (h)	To node: Node No. (i)	Blocks: Left (j)	Blocks: Right (k)	Left addresses: Low (l)	Left addresses: High (m)	Right addresses: Low (n)	Right addresses: High (o)
33	34 35 36	37 38 39 40	41 42 43	44 45 46 47	48 49 50	51 52 53	54 55 56 57 58	59 60 61 62 63	64 65 66 67 68	69 70 71 72 73
	3	43	3	44	305	410	37	37	35	35
9		↓		↓	↓	↓	38	44	46	46
9		↓		↓	↓	↓	48	48	39	49
		44		45	306	410	50	50	51	51
9		↓		↓	↓	↓	53	53	54	54
		45		46	306	409	57	59	61	63
9		↓		↓	↓	↓	60	62	64	64
		46		47	307	409	67	67	66	68
9		↓		↓	↓	↓	70	70	0	0
	↓	47	↓	48	307	408	72	76	73	77

Figure 6.9. Misaligned streets require additional nodes for six block faces (*top*). Downward arrows on the coding sheet (*bottom*) call for repeating the number in the rows below, and a 9 in the first column reflects an additional address range on the same block face.

Another facet of Corbett's framework invoked the dimensionality of objects by treating nodes, arcs, and polygons as 0-cells, 1-cells, and 2-cells, respectively. Because a node has neither length nor area, it's a zero-dimension object, whereas an arc has one dimension, its length, and a polygon is a two-dimensional object, with an area. Dimensionality also informs the mathematical notions of *bound* and *cobound*, which describe a cell's relationship to neighboring entities with a higher or lower dimension. In this schema, two 0-cells (nodes) on the same line bound the intervening 1-cell (arc) insofar as they limit its length, and the same 1-cell is cobounded by the pair of 2-cells (polygons) that it separates. Census Bureau data editing software confirms a DIME file's completeness and reliability by asking the questions:

1. **Is every 1-cell bounded by exactly two 0-cells?**
2. **Is every 1-cell cobounded by exactly two 2-cells?**
3. **Is every 2-cell bounded by a single closed, properly oriented sequence of 1-cells?**
4. **Is every 0-cell surrounded by a single closed chain of 2-cells cobounding the 1-cells linked to the 0-cell?**[31]

The first two questions address the simple requirement of bounding and cobounding: each 1-cell is a single DIME record with exactly two bounding 0-cells and two cobounding 2-cells. The third question tracks around the outside of each polygon and insists that the perimeter end where it started; in other words, every polygon must be completely bounded. If the perimeter does not close, or if a DIME record that mentions a 1-cell does not fit into a closed perimeter, the error must be corrected. The fourth question, which focuses on 0-cells, is a "dual" of question 3 that insists each 0-cell be surrounded by a closed sequence of abutting 2-cells. Complicated but useful.

Other tests for topological consistency confirm that a DIME file represents a smooth surface without discontinuities or tears. Marvin White, a colleague of Corbett's, showed how the mistaken assignment of the same number to two different nodes in the network can create a "handle" that violates the "planarity" requirement.[32] In figure 6.10, for example, node 23, which bounds four 1-cells cobounded by polygons *A*, *B*, *C*, and *D*, has been confused with the node surrounded by polygons *M*, *N*, *O*, and *P*. Although the consequences for census enumeration are difficult to fathom, this impossible situation must be identified and corrected.

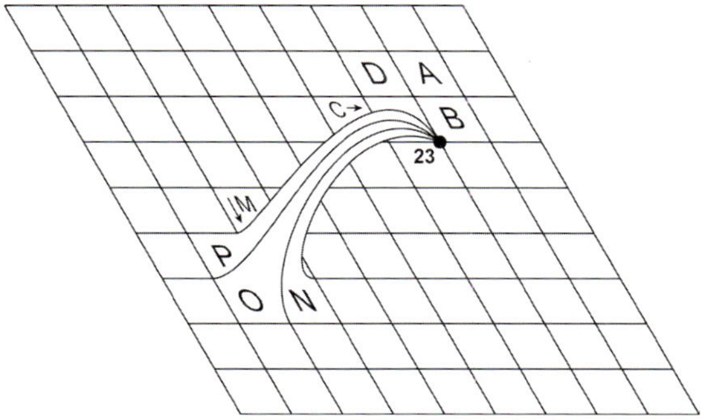

Figure 6.10. Topological screening can detect the "handle" that results when two nodes are assigned the same number, 23, which pulls 1-cells shared among the 2-cells *M*, *N*, *O*, and *P* over toward the real node 23.

Errors such as this, if undetected, would undermine the reliability of GPS navigation systems used daily by millions of cars and trucks. All satnavs feature what software engineers call the *two-point shortest-path query*: enter a destination, and within seconds, the system calculates the shortest, quickest, least expensive, or most scenic route from where you are to where you want to go. These searches depend on error-free topologically structured road data.

Although numerous shortest-path algorithms exist, most are a variation of the recipe devised by Edsger Dijkstra (1930–2002), a Dutch computer scientist who advanced the design of operating systems and computing networks.[33] According to his *New York Times* obituary, the idea for the algorithm "came to him one morning in 1956 as he sat sipping coffee on the terrace of an Amsterdam café."[34] The simple map in figure 6.11 illustrates his strategy for finding the optimum route from node *A* to node *H*—because the numbers along the links differ markedly from their geometric lengths, think of them as travel time, which satnav users typically want to minimize.

Dijkstra proposed a step-by-step approach that works outward from the starting point by visiting new nodes one link beyond those already visited in the previous round, calculating the minimum distance from the starting point and replacing a node's minimum distance whenever a shorter route is found. The number inside each circle is the tentative minimum distance to the node for the round indicated. To keep track of changes, he divided the nodes into three sets: nodes for which the

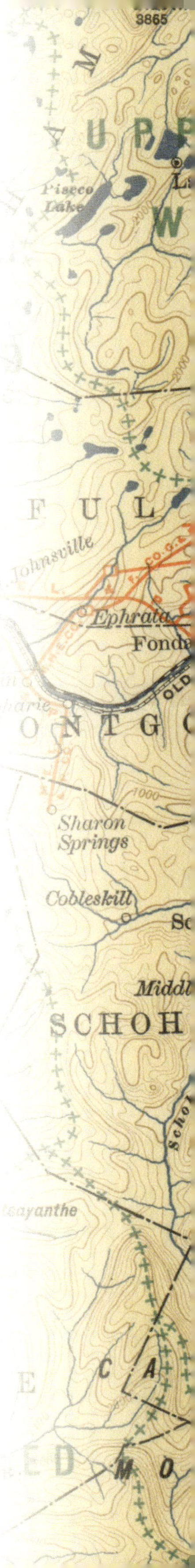

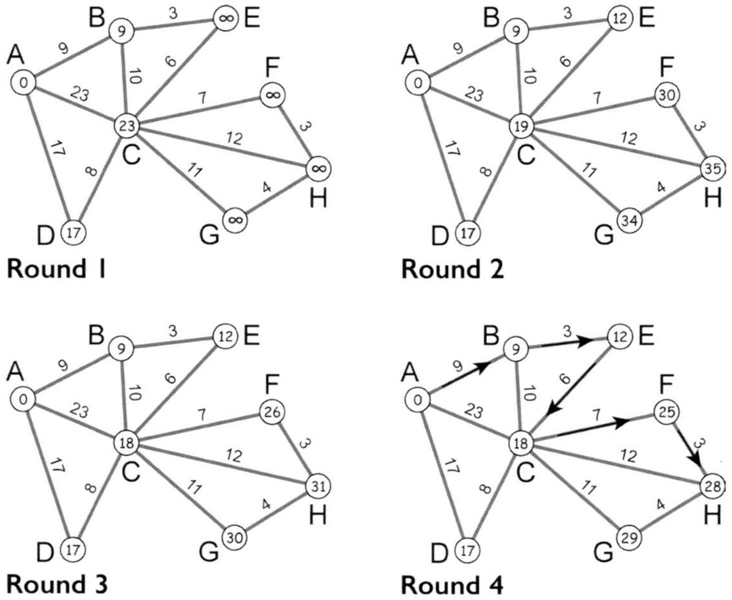

Figure 6.11. Successive rounds of the Dijkstra algorithm yield a minimum distance path from *A* to *H* via nodes *B*, *E*, *C*, and *F*. Numbers in the circles are the minimum distance from *A* at the end of the round.

minimum distance from starting point *A* is known, nodes for which a tentative minimum distance has been found, and nodes that have not yet been visited. The initial map (round 0) has a minimum distance of 0 for node *A* and infinity for all the other nodes. The map for round 1 (figure 6.11, upper left) merely shows the direct distance from *A* as the minimum distance to one-step nodes *B*, *C*, and *D*. Round 2 (figure 6.11, upper right) uses the round 1 minimums to extend the area visited out to nodes *E*, *F*, *G*, and *H*. The distance to *E* is set to 12 (9 + 3, through *B*), and the round 1 distance to *C* (23) yields tentative minimum distances for *F* (23 + 7 = 30), *G* (23 + 11 = 34), and *H* (23 + 12 = 35). In addition, because node *B* provides a shortcut between *A* and *C*, the minimum distance to *C* drops to 19.

Each new round begins by adjusting downward any tentative distances lowered in the preceding round. Hence, *C*'s new round 2 minimum (19) further lowers the minimum distances to *F*, *G*, and *H* (to 26, 30, and 31). There are no new nodes to visit in round 3, but discovery of a shortcut from *B* to *C* through *E* drops the minimum distance to *C* from 19 to 18. In round 4, this latter reduction lowers the minimums for *F* and *G* to 25 and 29, respectively, and a newly discovered shortcut through *F* reduces the minimum distance to *H* from 31 to 28. All nodes are now in the first set, for which minimum distance to *A* is known, and the process halts. The round 4 (figure 6.11, lower right) map shows the optimal route as *A-B-E-C-F-H*. Although the lengths of the links might make this path appear counterintuitive, most drivers are aware that road distance is not a consistently reliable predictor of travel time. For users seeking a scenic route, penalties can be added to links considered industrial, visually polluted, or otherwise tainted.

To support minimum-travel-time solutions, the physical lengths of the electronic map's links must be enhanced with attributes that reflect speed limits as well as the traffic impedance at intersections that have traffic lights or stop signs and the directional differences related to slope and congestion.[35] Uphill travel consumes more fuel, and perhaps more time, than downhill travel, after all, and a trip inbound toward the city center might take longer than an outbound trip, at least in the morning. Indeed, travel time usually varies with hour of the day, day of the week, and even month of the year in areas with seasonal tourism or disruptive summer road maintenance.[36] Some of these attributes are static and need not be updated instantaneously or even daily; static enhancements to specific links include one-way traffic, the posted speed limit, and toll collection, which can vary with direction—before "cashless" toll collection, bridges and tunnels linking New Jersey and New York were toll-free for westbound traffic but eastbound drivers faced congestion at toll booths.[37] For systems that provide turn-by-turn voice directions, landmarks such as gas stations or banks can be especially helpful.[38] Dynamic attributes, if the system supports them, include designated detours, advance reports of marathons and other scheduled events likely to frustrate drivers, and timely reports of accidents and road congestion.

Motorists are not the only beneficiaries. In cities where residents and visitors routinely walk or take public transportation, navigation apps on their mobile devices commonly mix different modes of travel, as when a journey that begins on foot includes the number 32 bus, the Green Line subway, and a transfer to the Blue Line, followed by a short walk to the destination. The city-wise app also shows when the next bus or subway train will arrive at a nearby stop—not the scheduled time but the actual time based on the conveyance's current location. And

travelers who opt for a ride-sharing service can monitor their driver's approach in real time on their mobile device.

Mobile and in-vehicle navigation apps depend on the real-time integration of their topologically structured data with the constellation of GPS satellites, which provides the coordinates needed to align map symbols with the flashing or otherwise prominent you-are-here dot at the center of the screen. Because the coordinates are estimates that might be off by several meters, "map-matching algorithms" instantaneously correct for positional error by snapping the user's location to the nearest network link—which explains the app's occasional momentary confusion when a trip begins near an intersection.[39] Mobile devices also require a wireless connection to a web map service, which sends whatever small individual map sections, known as *tiles*, are needed to update the display as the user moves around or zooms in or out.[40]

By contrast, built-in automotive navigation systems have commonly relied on map files stored internally and updated every year or so. Although the stored data obviate dynamic tiling over the wireless network, dashboard systems in cars and trucks are often equipped to receive "Traffic jam ahead!" alerts from a two-way wireless crowdsourcing service that monitors thousands of cars and trucks. When the congestion avoidance system determines that vehicles are moving sluggishly along a particular link, or perhaps not at all, it sends a digitally coded congestion report over the designated traffic messaging channel so that satnavs in affected vehicles can update their maps—but only for paid subscribers.[41] Car buyers given a trial subscription are easily hooked on a paid service that many consider as indispensable as satellite radio. But drivers who attach their mobile device to a dashboard or window mount can receive a similar service for free.[42] Or so it seems: our GPS traces are valuable data, and savvy users appreciate the cliché that if they're not paying, they're the product.[43]

Engineers who equate the automated sharing of traffic information to the synchronized behavior of ant colonies conceived the term "ant-based vehicle congestion avoidance system."[44] The prospect of self-driving cars suggests a further analogy—ant-based vehicle infestation—as when traffic congestion triggers invasions of residential neighborhoods by robot drivers following a detour suggested by their satnav's algorithmic instincts.

That unintended consequences can prompt unforeseen reactions became apparent in late 2017 after irate residents of Leonia, New Jersey, near the George Washington Bridge, convinced local officials to resist traffic diversions attributed to navigation apps such as Google Maps™ and Waze®.[45] "Do Not Enter" signs targeting rush-hour traffic ("6 a.m. to 10 a.m., 4 p.m. to 9 p.m.") carried a "Residents Exempt" addendum. To avoid traffic tickets, borough residents hung official yellow tags inside their windshields. Parents of young children were pleased, but business owners dismayed by lost sales pressured officials to modify the ban. One nearby municipality enacted its own ban, and another sued Leonia for violating its citizens' rights. Litigation and complaints persisted after New Jersey's attorney general declared the restrictions unenforceable.[46] Speed bumps might be the ultimate solution.

Waze, one of the mobile apps criticized in the Leonia traffic diversions, has been chastised for hampering traffic enforcement.[47] The ultimate in crowdsourced mobile navigation, it lets volunteers set or clear warnings about accidents and police activity such as speed traps and sobriety checks: a driver who notices a speed check that's not on the screen can add it, and if Waze displays a warning that is no longer relevant, the driver can click "Not there." The example of a Waze screen on the left side of figure 6.12 shows the route ahead along Merlin Avenue and a detour via St. Martin to avoid an accident and traffic congestion. Off to the left, a tiny face with a mustache and a police officer's peaked hat (in blue) with a shield above the visor signifies what might be a speed trap. Touch the large circle (in orange) near the lower-right corner of the screen, and a menu of alerts and services (figure 6.12, right) pops up. Some drivers can add Waze to the larger screen of their vehicle's infotainment system.[48] Warnings might eventually include red-light and speed cameras.[49]

Drivers not part of the Waze community can also benefit from crowdsourcing. Congestion avoidance systems that capture traffic flow data can enrich the map by detecting locations with stop signs, estimating average wait time at traffic lights, and using this information to calculate shortest-path routes more reliable than routes based only on static attributes such as distance, road category, speed limit, and number of traffic lights.[50] These enhanced network data can be licensed to navigation systems without dynamic extra-fee congestion avoidance, and more reliable paths calculated between key junctions can be saved in a "cache," from

Figure 6.12. Typical Waze screen on a mobile device (*left*) and the menu of alerts that the driver can add or clear, (*right*).

which results are recalled rapidly and sent out as needed whenever a motorist requests an optimal route.[51]

No less important than congestion history is a road's safety record, compiled by highway departments from accident reports by police, motorists, and insurance companies. Collected to identify roads that might need better signage, a lower speed limit, a design improvement, or an increased police presence, accident data can enhance shortest-path routing algorithms by penalizing stretches of highway with above-average accident or fatality rates.[52]

Specialized road data can also warn truck and bus drivers of weight restrictions on bridges as well as height and width clearances for underpasses and tunnels.[53] But getting drivers to pay attention might require a tattletale navigation system that not only warns drivers about road hazards but also reports violations automatically to the employer or local law enforcement, perhaps at the insistence of the firm's insurer. I can understand the frustration of localities that not only invest in electronic height monitors, flashing lights, and huge signs that prudently understate the maximum height but also paint bright-orange stripes on the sides of low bridges that repeatedly rip the tops off box trucks and tractor trailers.

Sometimes it's a bus, and people die, as happened one night in September 2010 at 2:30 a.m. a few miles from where I live when a 13-foot-tall double-decker intercity bus hit a railroad bridge with a posted clearance of 10 ft., 9 in. The bus driver, who had lost his way, was looking at his personal satnav when he hit the bridge at 35 mph, killing four passengers and injuring at least a dozen others.[54] Not noted as a road hazard on satnavs, the bridge continues to make news several times a year, mostly because of trucks from out of town. But a few are rented trucks driven by local do-it-yourself movers not accustomed to overhead clearance as a hazard.

Railroad grade crossings warrant special attention, particularly crossings with a track-high hump that can snag car carriers, lowboy equipment trailers, and other vehicles with little ground clearance or a long wheelbase. A customized navigation system that incorporates the vehicle's specific vulnerability, including its cargo, might warn truckers away from routes with a high probability of getting T-boned by a train. Grade crossings not protected by a crossing gate are especially hazardous at night or where a curved right-of-way or trackside structures limit visibility for both the engineer and the trucker. A satnav that provides both visual and audible warnings about grade crossings might even include a wireless reporting system like Waze's touch-screen alerts so that truckers can promptly report malfunctioning crossing gates or signals to the railway company and highway officials. Safety could be further enhanced by a trucker's navigation system able to detect and report when it's stalled on the tracks with a load of propane or crushed stone. Interconnected navigation systems can be an important safety measure when transport networks intersect.

Railroad safety demands not only GPS units that monitor a train's speed and location but also intelligent control systems that use electronic maps to prevent trains from running into each other, hitting a misaligned switch, or derailing on a curve because of excessive speed. Known as positive train control (PTC), these integrated systems are a technological and political response to costly accidents, many fatal, and mostly attributed to distracted engineers, excessive speed, inattentive dispatching, massive objects on the track, and cost-control strategies such as deferred maintenance and one-person locomotive crews.[55]

A good example of the need for PTC is Amtrak train 188, which was traveling at 106 mph when it derailed in Philadelphia on May 12, 2015, on a curve with a 50 mph speed limit, killing eight passengers and injuring nearly two hundred others.[56] A National Transportation Safety Board panel concluded that the engineer, who was awake and not under the influence of drugs or alcohol, believed he was in a 110 mph zone. Despite the engineer's "loss of situational awareness," the accident could have been prevented "if positive train control or another control system had been in place to enforce the permanent speed restriction of 50 mph at the Franklin Junction curve."[57]

As described in figure 6.13, from the Federal Railroad Administration's Office of Research and Development, PTC relies on radio communication to integrate an electronic map of track and switches at headquarters with GPS receivers in locomotives, GPS receivers in maintenance-of-way and track-inspection vehicles, the dispatcher's control system and map, and various "wayside interface units," such as train defect detectors and signals at rail and highway crossings.[58] Had Amtrak's PTC been fully implemented at the time, it would have prevented the engineer from increasing speed as he approached the curve—yes, he actually accelerated from 95 mph on the straightaway before the curve.

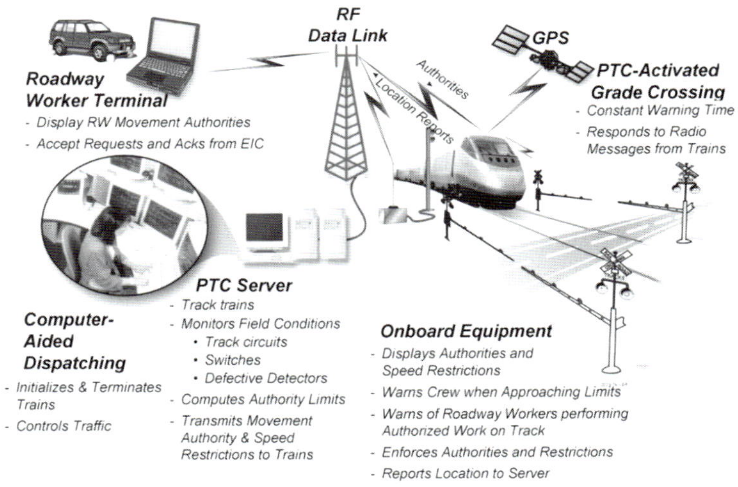

Figure 6.13. System architecture of PTC.

PTC transfers much of the responsibility for situation awareness from humans to computers. It is several steps above and beyond steam-era technology such as the *dead man's switch*, which shut a train down when an engineer who had a stroke or heart attack took his hand off the throttle, or automatic *block signaling*, whereby visual trackside or in-cab warnings told an engineer not to enter a section of track occupied by another train. Congress told the railroads in 2008 to implement PTC by 2015, but many lines responded slowly, and in October 2015, the government extended the compliance deadline to the end of 2018 with the possibility of a further extension to the end of 2020 for railroads that had made substantial progress.[59]

Although PTC exemplifies the power of electronic maps and location monitoring to orchestrate the safe and efficient movement of thousands of trains through the national rail net, topology has an important role in identifying efficient routes provided by passing sidings and multiple-track routes with crossovers. Dynamic routing of passenger trains around slow freights can alleviate the frustration of long-distance passengers otherwise forced to tag along behind a commuter local or sluggish freight, as well as help shepherd trains through labyrinths of trackage at rail yards and large passenger terminals. Similarly, electronic guidance and topological road data might significantly reduce energy consumption while orchestrating a safe realization of driverless cars and trucks. Control has been a persistent theme in map history, and with the growing public concern about climate change, energy consumption, and traffic congestion, the map's role is understandably expanding from control of territory to control of movement—no doubt with unintended consequences yet to emerge.

7

CONTROL

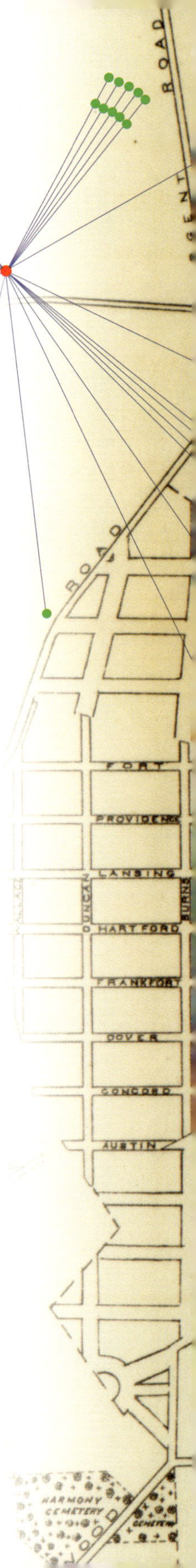

In reworking the outline for this book, I could find no better single-word theme for this concluding chapter than *control*, which coincidentally emerged as the capstone concept in the previous chapter, in which networks provided a framework for governing the movement of wheeled vehicles through webs of highways and tracks. As a chapter title, "control" is a conveniently ambiguous concept, useful because its multiple meanings foster diverse vignettes, individually fascinating (to me, at least) and collectively relevant to an integrated exploration of maps and networks. Their diversity provides an open-endedness that I hope will encourage readers to reflect on the notion that behind every great map is a network and behind every great network is a map.

This chapter looks at map/network control from several directions, starting with motor vehicles, trains, and unmanned aerial vehicles (UAVs, aka *drones*). Discussion then moves to the dramatic impacts of the internet and social media, which have extended both the reach and the grasp of telecommunications in unprecedented ways, and then to a concise examination of the transformative effect of topologically organized geographic data on electoral redistricting and American democracy. These vignettes bolster the argument that maps and networks have converged markedly in both practice and concept since topological databases began to pervade mapping and map use in the late 1960s. Manipulation of maps for political purposes is the height of cartographic control.

Ambiguity of a different sort is apparent in the collision-avoidance systems proposed for driverless cars and delivery drones. One strategy calls for costly new infrastructure, such as detection circuits embedded in streets and walkways, while another would rely on a combination of machine vision robotics and digital maps.[1] Although existing electronic maps are generally adequate for automobiles—assuming conscientious updating—drones would require additional data describing vertical obstructions such as utility lines and tall buildings as well as the horizontal and vertical flight restrictions imposed by federal and local regulators and the locations of charging or battery-swap stations. Knowing what to avoid is as important as knowing where to go.

As the previous chapter noted, transportation networks stored electronically for analysis and display offer motorists a continually updated view of the immediate area and the road ahead. The map on the screen appears convincing because its software uses the comparatively reliable coordinates of the digital road network to instantaneously correct a GPS estimate accurate, at best, to no more than several meters.[2] As an exemplar of cross-network synergy, the satnav snaps the display's you-are-here dot onto the vetted and more precise road network, thereby adding value to the costly constellation of GPS satellites and its network of ground stations, originally intended to guide cruise missiles. Additional ground-based reference stations installed for the Wide Area Augmentation System (WAAS), a corrective strategy sponsored by the Federal Aviation Administration (FAA) and the US Department of Transportation (DOT), make satellite positioning sufficiently reliable for commercial aviation and drones.[3]

The enduring military imperative of command and control pervades vehicle navigation systems, which not only calculate optimum itineraries but insist that their directions be obeyed. Although we can specify the destination and minimize distance or travel time while avoiding tolls, seeking scenery, or dodging left turns, many navigation systems nag incessantly whenever we forsake their prescribed path for a known shortcut or an impulsive side trip. Intentional departures are often met with annoying pleas to rejoin the programmed route. Because my car's built-in satnav lacks a "pause" button, I typically endure multiple rounds of "recalculating" until I press Cancel or the device detects a suitably short alternative aligned with my ad hoc detour. I'm hoping that more intelligent satnavs might eventually figure out an individual driver's deference to curiosity, local knowledge, or mandatory diversions around construction projects or fire apparatus.

Although owners of driverless cars might willingly cede the entire route to a robot chauffeur who promises efficient hands-and-eyes-free delivery to the chosen destination, I can imagine a system that lets the user insert waypoints, perhaps with "a check this out" option for exploring someone else's local favorites. Social Media meets the Driverless Car? Will autonomous cars have an optional steering wheel—no doubt with restrictions on when, where, and how it can be used? The collective benefits of driverless cars might depend on minimizing interaction with less-disciplined and unpredictable human motorists. Driverless cars will work largely because they can talk to each other.

Other vehicles are not the only problem. Machine vision algorithms for identifying pedestrians and roadside objects raise ethical concerns about who dies when a collision is unavoidable.[4] Should the vehicle hit a young person who is jaywalking or crash into a bridge abutment and kill its sole occupant? What if the jaywalking pedestrian is holding a child? What if the vehicle has three occupants? What if all are in their 80s? Although social scientists and philosophers might craft acceptable algorithms, they must also consider the reliability of unaided machine vision algorithms, which could be better informed by an electronic map with the locations of crosswalks and solid objects. A well-informed network could promote safety by forcing the autonomous vehicle to slow down and be more observant near locations where collisions have occurred or where pedestrians are known to step into the road.

Who-dies scenarios aside, benefits would arise from accommodating more traffic with fewer, narrower roads, in some cases by joining multiple vehicles into trainlike platoons. Because pooled wind resistance can reduce fuel cost, this option seems a plausible strategy when origins and destination are shared over distances sufficiently long to offset delays incurred in assembling and disbanding the convoy. Platooning seems particularly beneficial for trucks, but look for resistance from truck drivers faced with job losses and from railway companies, which provide their own track and signaling systems.[5] Of course, the heavily monitored system of superhighways needed for truck platoons would simplify the collection of fees necessary to maintain the system, recover investment cost, and replace the gasoline tax revenue lost when electric motors make internal combustion engines obsolete.

Autonomous vehicles are an inevitable next step. For several decades, computerized traffic management systems have helped highway administrators in large metropolitan areas improve, or at least cope with, traffic flow on freeways and city

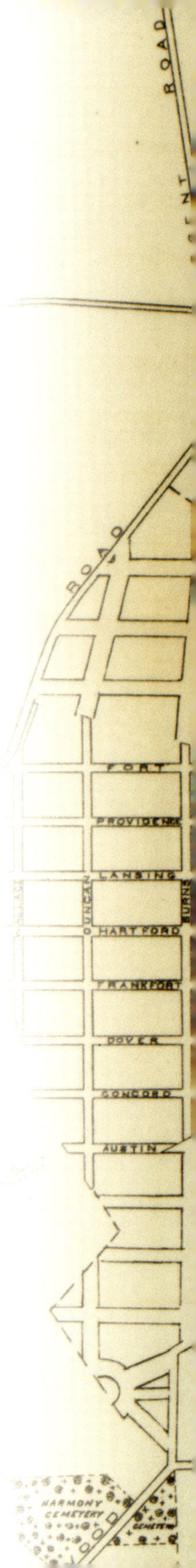

streets. Typical approaches monitor the rate of traffic flowing past critical points, synchronize traffic lights to avoid wasted capacity, and carefully regulate vehicles merging onto a freeway at each on-ramp. Traffic management systems depend on electronic maps of the road network and queuing software able to predict congestion and manipulate traffic signals to minimize backups.[6] With or without machine vision, electronic cartography will eventually oversee and micromanage where and when we go.

Interim strategies might include engine governors to enforce speed limits and "black box" computers to help insurers identify distracted, reckless, or aggressive drivers. A satnav integrated with a recorder becomes a vigilant snitch useful to both law enforcement and the auto insurance industry. GPS units typically incorporate an accelerometer, which makes them a better speedometer than legacy equipment that merely times the rotation of the front wheels. In addition to comparing a vehicle's actual speed with the posted speed limit, a driving monitor can detect and record sudden lane changes, frequent braking, and other symptoms of dangerous driving.[7] To address privacy concerns, insurers can offer lower rates to motorists who accept GPS monitoring, and courts could make this a requirement for motorists prone to collisions or moving violations. Many cars have tracking devices embedded in their electronics, added by manufacturers leery of lawsuits over accidents that could be blamed on poor design rather than bad driving.[8] Should a car's computer be allowed to testify against its driver?

Privacy is hardly an issue for positive train control, which can enforce speed restrictions and prevent derailments and rear-end collisions. Indeed, railroads that resist safety systems mandated by statute are arguably liable for whatever accidents PTC might have prevented. By contrast, a train company might protect its network by installing sensors at grade crossings to warn of vehicles stalled on the track or by flying an inspection drone a mile or so ahead of the locomotive to watch for landslides and broken rails as well as stalled vehicles.

Although protecting every train with an aerial scout is hardly practicable—on many days gusty winds would preclude flying—some railroads are using drone surveillance to check for hazardous conditions, such as eroded roadbed, inundated track, avalanches, and oil or rocks on the rails.[9] GPS, an electronic map of the network, and the roadbed itself can keep a UAV on course, and the database can pinpoint sections of track requiring close scrutiny at a slower speed. Although

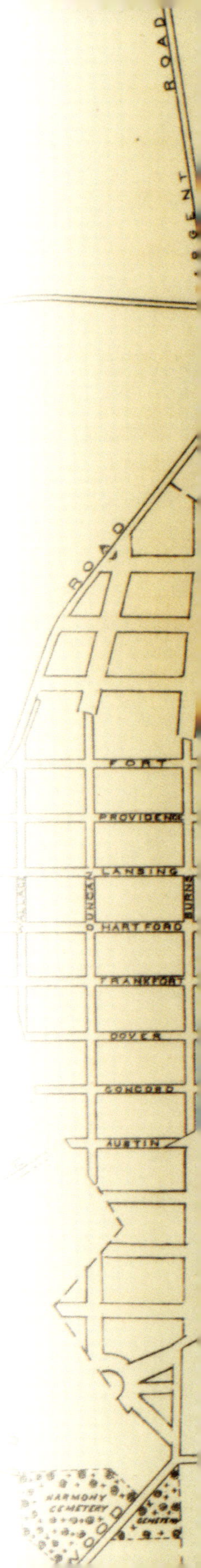

conventional ultrasound systems for detecting flaws in steel rails are too heavy for inspection drones, comparatively compact lidar systems might eventually screen for hairline cracks. In addition to routine scheduled monitoring, aerial inspection of the track might be particularly appropriate after a heavy rain or noteworthy earth tremor.

Drones can also help electrical utilities manage their extensive networks of power lines, transmission towers, and other elevated facilities not easily observed from the ground. For systematic aerial surveillance of transmission lines along a cleared right-of-way, UAVs offer a safer and more cost-effective substitute for manned helicopters.[10] They also allow overhead examination of fields of solar panels as well as the network of distribution lines, closer to the ground and aligned with the street network. In residential neighborhoods and rural areas, drones with cameras can help assess storm damage as well as promote efficient tree trimming, which depends on visual inspection of vegetation likely to encroach on wires during a windstorm or after a snowfall. In the near run at least, drone inspections are likely to be under the direct control of maintenance crews on the ground, as a supplement to the bucket trucks used in working on overhead lines and transformers.

UAVs guided by electronic street maps can help municipalities better meet their surveillance obligations. Police departments, for instance, can use drones to supplement or substitute for road patrols by screening for suspicious persons or vehicles. And watchful highway departments can deploy specially designed UAVs to look for cracked pavement or hazardous accumulations of snow, leaves, or water: dangers less likely to be reported by citizens upset by potholes or backed-up storm sewers, which systematic vigilance can avoid. While police drones with conventional digital cameras might stir up angst about civil liberties, highway inspection drones with specially configured imaging systems might also raise concern about their cost and reliability.[11]

More ambitious proposals for street-wise UAVs hypothesize a commercial pickup-and-delivery service faster but not greatly different from what's offered by the US Postal Service and its commercial competitors. The emergence of Amazon® as the country's most prominent online retailer led to proposals for personalized delivery by GPS-guided drones dispatched from local warehouses. Though the need, economics, and reliability of this level of customization are questionable,

Amazon's field tests triggered serious discussion by conventional parcel services, marketing experts, and local officials fearing an invasion of buzzing couriers. In addition to the noise and visual distraction, delivery drones pose real or perceived threats to personal privacy—Is that quadcopter carrying a package to someone's doorstep, or is its cargo a camera collecting quick peeks into our bedrooms and bathrooms? At a minimum, delivery drones would need to be licensed. Stormy weather is another impediment.

If delivery drones ever take hold, electronic cartography could assume the dual roles of enabler and regulator, guiding Amazon Prime® UAVs along the public right-of-way above the heads of pedestrians and allowing brief access trips across the recipient's front yard but otherwise banning access to quiet streets from late evening through early morning—an add-on of sorts to existing noise and nuisance ordinances that restrict loud construction equipment to normal waking hours and discourage motorists from taking shortcuts through residential neighborhoods. For delivery drones, specialized aeronautical charts would resemble zoning maps, configured to preserve quality of life and sustain property values.[12] In addition, public pressure could sanctify school grounds and cemeteries as no-fly zones similar to public-safety restrictions on hovering over sports stadiums, amusement parks, and other targets attractive to terrorists or mischief-makers.

Can local agencies regulate drones within their borders? Of course. Although the FAA has claimed the sole right to regulate nonmilitary aircraft, local police, traffic officials, and zoning boards understandably see low-altitude UAVs as more akin to cars and trucks than to conventional planes and helicopters. Whatever three-dimensional boundaries carve out local exceptions to the national airspace will empower a vast series of large-scale aeronautical charts that rely on GPS and highly precise electronic maps to control where and when drones can fly and what they can do. These state or local UAV charts will be necessary even if delivery drones never become much more than a boutique parcel service.

As commercial aviation demonstrated, new networked technologies demand new rules, and these rules typically require maps, exemplified in the early twentieth century by the aeronautical chart, which emerged as a distinctive genre of prohibitive cartography. It's not yet clear what maps might be needed to regulate the internet, largely a twenty-first-century phenomenon, but its geographic impacts have inspired their own cartography, readily available to anyone who Googles "map of the internet." Because both the cartographic history and the history of the network fascinate me, I'll start in the late 1960s with the packet-switching network developed at the Advanced Research Projects Agency (ARPA), a special Defense Department branch focused on aviation, spy satellites, and information technology and later named DARPA.

If a simple hand-drawn map didn't inspire the process of packet switching, maps made it easy to explain and sell the concept. As illustrated by an unidentified employee's pencil sketch (figure 7.1), the pioneer Advanced Research Projects Agency Network (ARPANET) used high-capacity long-distance telephone lines leased from AT&T® to link small computers known as IMPs (interface message processors) at four university research centers (the University of California, Los Angeles [UCLA], the University of California, Santa Barbara [UCSB], Stanford Research Institute, and the University of Utah), each with a host computer (Sigma 7, SDS 940, IBM 360, and PDP-10, respectively) linked to terminals on campus.[13] Messages and datasets intended for a particular destination were broken into small packets, each with a sequence number and the destination's address. The IMP then sent the packets over the network to one or more neighboring processors, and if the receiving IMP was not the intended destination, it forwarded the packet onward. For example, some parts of a message originating at UCSB might go to UCLA and the others to Stanford, and if the message was intended for Utah, some packets might pass only through Stanford while others transited UCLA and then Stanford. The IMP at the intended destination would string the packets together in the correct sequence and forward the complete message to the local host, which would hold it until the recipient logged in. With proper permissions, a user could log into a host computer at another site.

Packet switching allowed the network to grow by merely adding a single high-speed phone link to each new node. Even so, redundant paths were important because a link or node might be temporarily out of service, corrupted by static, or destroyed in an enemy attack. The military, which was funding development, sought a durable communication network able to function if some of its nodes and links were destroyed. Redundancy was apparent in the eastward extension of the network, which by late 1970 boasted two direct transcontinental links: one linking Utah and the Massachusetts Institute of Technology (MIT) and a second connecting the RAND Corporation (a think tank in Los Angeles) with BBN

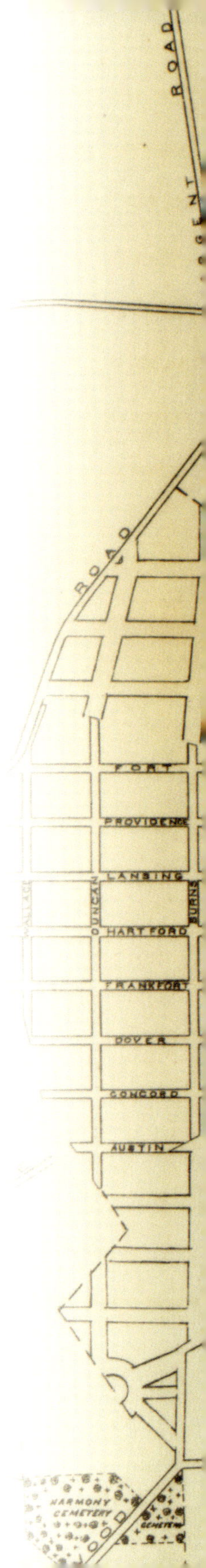

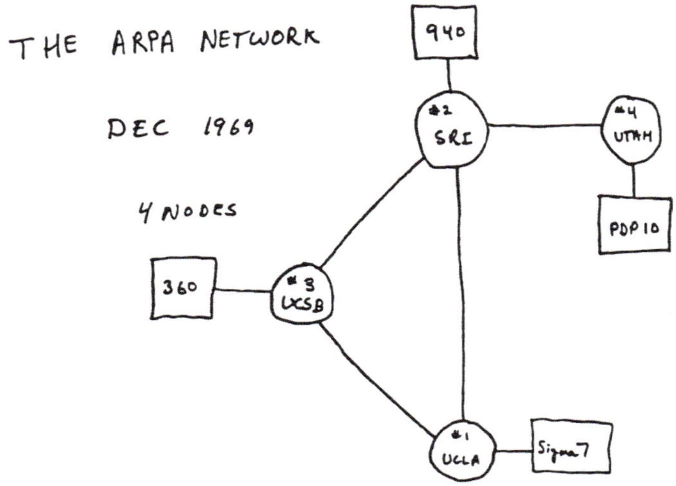

Figure 7.1. The original internet had only four nodes, all out west.

(Bolt Beranek and Newman, a research firm in Cambridge, Massachusetts).[14] By summer 1975, a densified network boasted a dozen nodes around San Francisco Bay and nine in Southern California, with a satellite link connecting NASA's Ames Research Center in Silicon Valley with Hawaii.

According to schematic diagrams that ARPA called "logical maps" (figure 7.2), some nodes handled packets with an IMP and others used a terminal interface processor (TIP), essentially an IMP enhanced with a small computer to service multiple terminals. Because the two different kinds of nodes made the display more visually complex, DARPA—the *D* for "Defense" was added in 1972—helped the public make better sense of its network by showing only its nodes and links on "geographic maps" such as figure 7.3. Circles that suggest magnifying glasses underscore the emergence of two clusters in the East: one in the Boston area, where Harvard and the Defense Department's Lincoln Laboratory (run by MIT) bracketed a string of three-letter research centers, and another near Washington, DC, that included the National Security Agency, Aberdeen Proving Ground, and Fort Belvoir, as well as a satellite link to the Norwegian Seismic Array (NORSAR). Established to detect earthquakes and violations of the Nuclear Test Ban Treaty, NORSAR provided a convenient European foothold for a link to London. Although the ARPANET was strongly bicoastal, its transcontinental links joined the network to important research centers in the heartland, notably Argonne National Laboratory, a Department of Energy facility outside Chicago; the Department of Commerce Boulder Laboratories (DOCB) in Colorado, which covered a broad range of physical sciences, including atmospheric research; and the Air Force Weapons Laboratory (AFWL), at Kirtland Air Force Base, near Albuquerque, New Mexico, dedicated to nuclear weaponry and defensive measures.

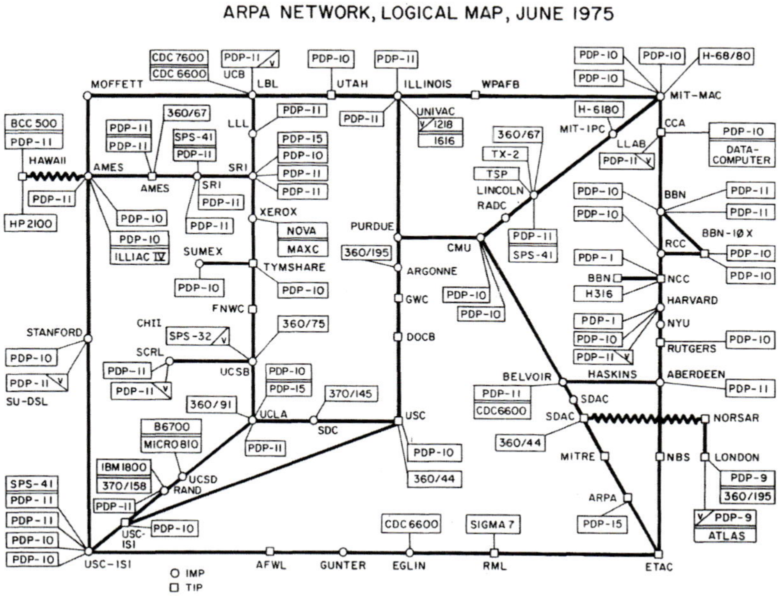

Figure 7.2. In June 1975, the ARPANET was dominated by government research facilities and universities.

Advances in networking technology in the 1970s stimulated an explosive growth and reconfiguration of computing networks in the 1980s.[15] DARPA, in keeping with its mission of cutting-edge research, transferred operational responsibilities for the ARPANET to the Defense Communications Agency in 1975, and the military moved its operational communications from the ARPANET to the Military Network (MILNET) in 1983.[16] Other networks emerged in the early 1980s, most notably the Computer Science Network (CSNET), set up by the National Science Foundation (NSF) to serve computer scientists in academia and industry, and the Because It's Time Network (BITNET), founded by a

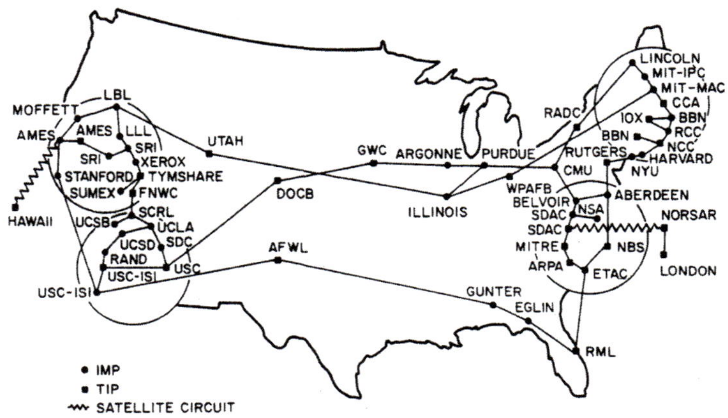

Figure 7.3. In July 1975, the ARPANET was strongly bicoastal, with three transcontinental links.

consortium of universities to serve academics in other fields.[17] In 1986, the NSF launched the National Science Foundation Network (NSFNET) to give academic researchers access to a half dozen federally funded supercomputers through a three-tier system consisting of a sparse, high-speed "backbone" connected to a larger number of important nodes by a midlevel of regional links, while a lower, more local tier served lesser nodes.[18]

Operational by mid-1988, the initial NSFNET (figure 7.4, top) resembled the ARPANET's configuration a dozen years earlier, hardly surprising because the NSF had essentially taken over ARPA's offspring. Nodes that appear to lie offshore merely reflect the cartographic license of an NSF staff artist who drafted the illustrations for an NSF history website.[19] Off the northeast coast, an arc of nodes still extends southward from MIT to Harvard; off the coast of the Carolinas, a group of nodes includes Aberdeen Proving Ground and the Pentagon; and reaching eastward toward Europe is a connection through NORSAR to London. The mapmaker's set of four maps includes a similarly disjoint rendering of the ARPANET's geographic map in figure 7.3. More cartographically correct are the heavy lines connecting the six supercomputers.

Although supercomputers proved nowhere as transformative as the NSF had hoped, by 1995 the NSFNET had acquired a markedly stronger backbone (figure 7.4, lower), labeled T3 to reflect 44 Mbps (megabits per second), nearly thirty times the 1.5 Mbps speed of the T1 backbone connections in 1988.[20] The

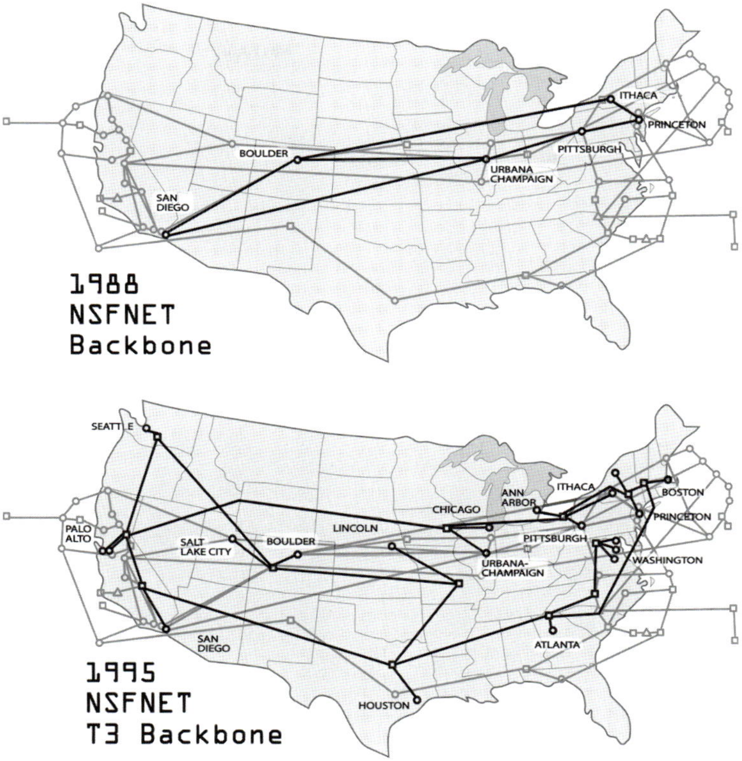

Figure 7.4. Maps of the NSFNET in 1988 (*top*) and 1995 (*bottom*) by NSF mapmaker Zina Deretsky.

T1 bandwidth dampened frequent complaints of university researchers frustrated by slow service, yet users quickly learned to expect ever higher capacities. Having demonstrated the power and prospects of networked computers, the government phased out the NSFNET in 1995 and transferred the traffic to the commercial networks that provided the fiber-optic cables, routers, and other infrastructure for what came to be called the *internet*.

Repeated attempts to map the internet attest to its prominence and wide reach.[21] These representations typically reflect the fundamental difference between figures 7.2 and 7.3, the first focusing on the (topo)logical links between sites and the second attempting a more geographically accurate depiction of location. As linkages and routings proliferated, some maps of *cyberspace* (as it came to be called) suppressed the connections altogether, to minimize graphic clutter, while others asserted a busy, often overwhelming connectedness, as exemplified in artist Zina Deretsky's map (figure 7.5) for the National Science Foundation's Where Discoveries Begin website. Titled *NSF and the Birth of the Internet*, her mildly rhetorical map touts transcontinental connections "between routers in major urban areas" that reflect the internet's growth "from humble beginnings in the academic research community .... as private-sector providers scrambled to meet the rising public demand for greater access and bandwidth."[22] In this context, most cartographic purists would excuse the rounded coastlines and urban hubs displaced from the Atlantic shoreline to southern Vermont and central Maine.

Figure 7.5. National Science Foundation artist Zina Deretsky, who created this portrait of the internet in 2007, credited maps by Chris Harrison, at Carnegie Mellon University's Human-Computer Interaction Institute.

Today's internet extends well beyond scientific telecommunications. The so-called "Internet of Things" not only links computers and mobile phones to alarm clocks, coffeemakers, thermostats, and security alarms but also governs a wide range of monitoring and triggering devices in applications as diverse as personal health, building security, food preparation, transportation, manufacturing, energy management, and environmental engineering.[23]

An application that stands out because of its links to both DARPA and cartographic representation is the SIGMA project, a dynamic network of fixed-location, vehicle-mounted, and wearable radiation detectors.[24] Figure 7.6 shows the instantaneous locations of volunteers with backpacks carrying mobile radiation monitors, walking about "in a scavenger hunt" for a fictitious "geneticist who had been mysteriously abducted."[25] The numbers indicate the number of sensors, and large numbers such as 65 and 107 reflect volunteers walking about in groups.[26] Designed originally to warn of dirty bombs, the program has been expanded to search for minute amounts of biological, chemical, and high-yield explosives.

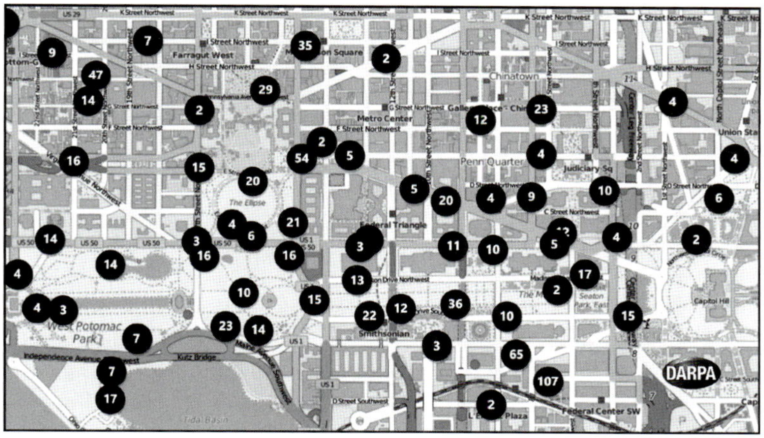

Figure 7.6. Snapshot of mobile radiation detectors during a 2016 test of the SIGMA threat warning system at the National Mall in Washington, DC. Dots represent individuals or groups of volunteers, and numbers inside the dots denote the number of sensors in their backpacks.

No less ominous is the internet's role as a conduit for social media, which had a troubling impact on the 2016 United States presidential election. Many voters apparently form their perceptions of politicians and national priorities from postings on Facebook and Twitter, which allowed trolls paid by the Russian government to tarnish the appeal of the Democrats' presidential candidate, Hillary Clinton. Clinton was strongly disliked by Russian president Vladimir Putin, who apparently believed that pejorative posts and forwarded fabrications would benefit her opponent, Donald Trump, who won the election even though he received 2.9 million fewer votes.[27] Clinton had a clear popular majority, but Trump received seventy-seven more electoral votes: the ones that matter in presidential politics. In the Electoral College, the candidate who wins a state's popular vote receives all its electoral votes, no matter how small his or her popular majority. Trump won because his 304 electoral votes trumped Clinton's 227 (figure 7.7).[28]

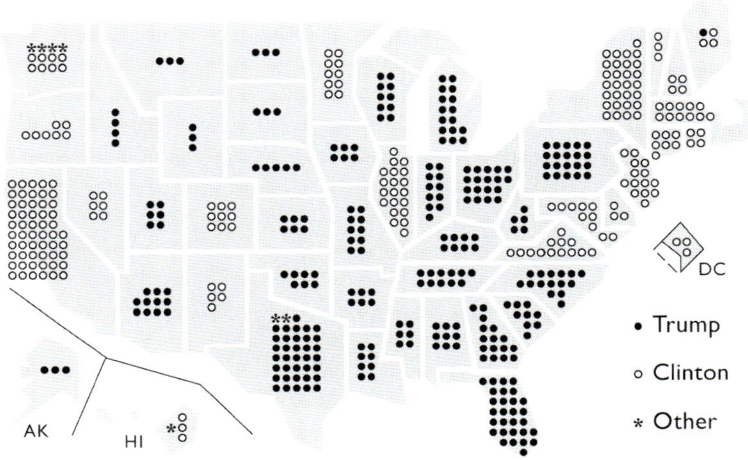

Figure 7.7. Electoral votes cast in the 2016 presidential election. Maine's four votes were split: three for Hilary Clinton, one for Donald Trump. Electors chose to vote for other people in Hawaii, Texas, and Washington.

This happened partly because our state boundaries are a structural *gerrymander*, an apt term for election districts that give a political party an advantage solely because of winner-take-all voting. The structural part reflects nearly two and a half centuries of historical happenstance and quirky political cartography: a complex narrative not readily dismissed as unfair or manipulative—it's there, it could benefit either party, and it's not easily changed. The gerrymander part reflects geographic concentrations of the losing party's voters into districts that deny it the victory that would result if there were no winner-take-all election districts. In a sense, Clinton lost because her voters were concentrated in too few states.

A seemingly straightforward proposal for mollifying this structural gerrymander would prorate electoral votes according to the popular vote—well within the scope of the Constitution. Maine and Nebraska do this, or try to, by awarding two electoral votes to the winner of the statewide popular vote and one electoral vote to the winner in each congressional district.[29] (A state has one electoral vote for each of its two senators, and one for each of its congressional districts, apportioned nationally according to population but too easily gerrymandered to serve a particular party.) In 2016, Trump won one of Maine's four votes and all of Nebraska's five.

To show how what's called the "Congressional District System" might affect the national tally, I drafted a hypothetical 2016 electoral map (figure 7.8) that reveals noteworthy shifts, particularly in Texas and Florida, where Clinton would have picked up fourteen and thirteen electoral votes, respectively, as well as in Minnesota, where Trump would have captured half the state's ten electoral votes, and in Virginia, where he would have received six of the state's thirteen votes.[30] In addition, Trump would have gained ground in big states such as New York, Illinois, California, and New Jersey, where he had multiple pockets of strength, while support for Clinton in cities and affluent suburbs would have undercut his winner-take-all victories in Pennsylvania, Michigan, and Ohio. Even so, in the final count, Trump would have won, 290–248 in electoral votes, because partisan gerrymandering concentrated Clinton voters in too few congressional districts. Although my map is hypothetical, Republican control of the House after the 2016 elections helped Trump fulfill his promise of a massive tax cut.[31]

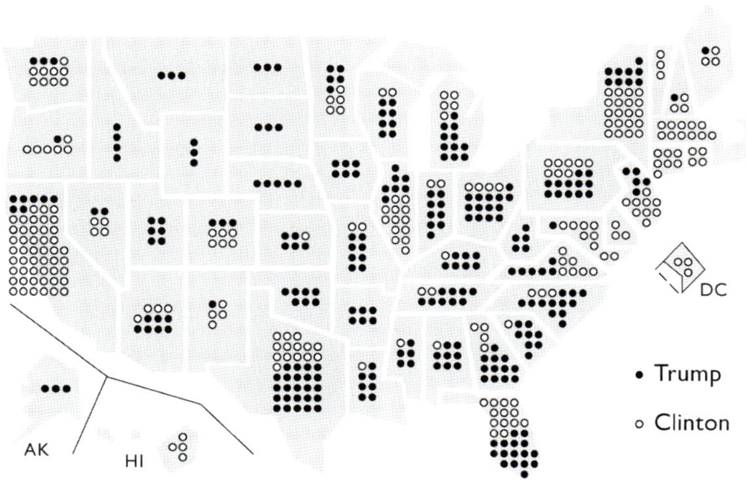

Figure 7.8. Electoral College results if 2016 presidential ballots were tallied according to the congressional district method.

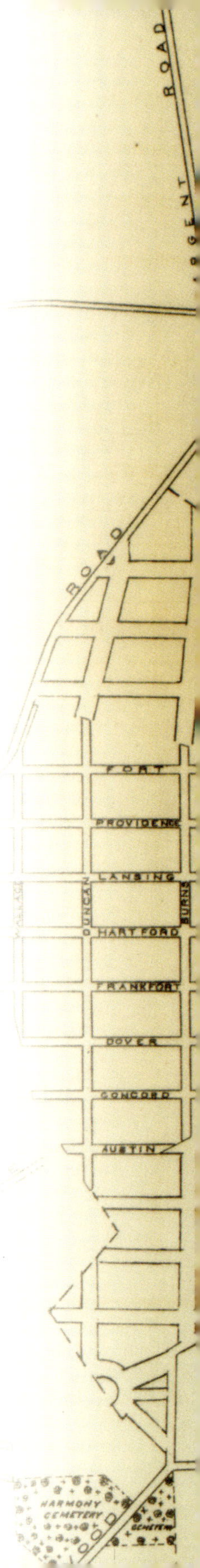

Partisan gerrymandering is the conscious manipulation of election district boundaries by the party in power, eager to increase its clout in both the state legislature and the House of Representatives. In most states, the party that controls the legislature controls redistricting, which becomes a game of packing voters likely to prefer the other party's candidates into relatively few districts. Play begins shortly after the Census Bureau fulfills its responsibility of counting heads every ten years, principally to reapportion House seats among the states. Both parties play this game, but in the 2010s the Republicans controlled more state legislatures than the Democrats, who saw their voters shoehorned into fewer districts.

Reapportionment is different from *redistricting*. If a state's population grew more rapidly than the nation's as a whole, it might gain one or more seats in the House. If it lost population or its growth was relatively sluggish, a state might lose a seat or two. But even if no seats are added or lost, some counties, cities, or neighborhoods now have proportionately more or fewer residents, and congressional boundaries must be adjusted to comply with Supreme Court decisions, starting in the 1960s, that ratcheted downward the acceptable disparity between a state's most- and least-populous districts. In the 1950s, South Dakota, which had 493,000 people in one House district and 159,000 in the other, led the nation with a disparity ratio of 3.1: an outrageous affront to the "one person, one vote" principle. Widespread violations of the Fourteenth Amendment's "equal protection" clause drew the scrutiny of the Supreme Court, which became increasingly less tolerant of numerical imbalance within states.[32] In 1969, the justices ruled against a New York map with a disparity ratio of 1.139, and in 1973 and 1983, respectively, they struck down redistricting plans with ratios of 1.042 and 1.007 in Texas and New Jersey.

Network cartography became important after the Census Bureau published block-level population counts and redistricting officials learned to use topologically structured data to satisfy the equal-population mandate while pursuing partisan objectives. Although the 1940 Census provided block counts for some areas, nationwide coverage was not available until the 1990 tabulations, which coincided with the release of TIGER/Line® street-network data. Meanwhile, amendments to the Voting Rights Act inspired the Department of Justice to pressure state legislatures to increase the number of districts in which blacks or Latinos constituted a "supermajority" sufficiently decisive to elect one of their own.[33]

Improved computers, efficient redistricting software, and clever politicians crafted "minority-majority" districts with bizarre shapes such as those in figure 7.9, ridiculed in the media with labels like "Pair of Earmuffs" (Illinois Fourth) and "Bullwinkle [the Moose]" (New York Twelfth).[34] Although a 1993 Supreme Court decision rejected the use of race as the dominant criterion, partisan gerrymandering in which race was not the sole factor remained constitutional.[35]

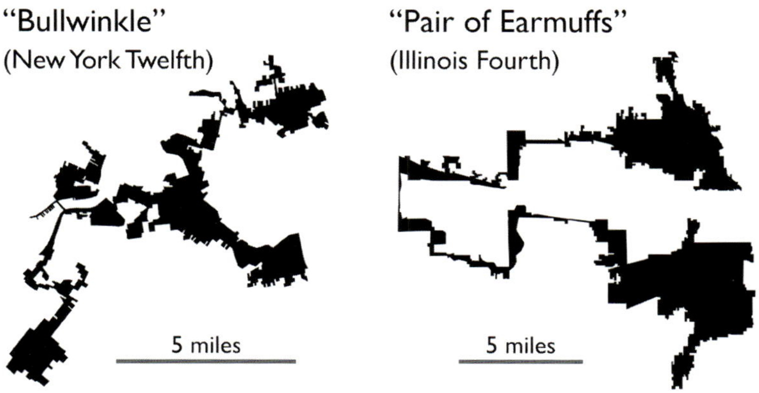

Figure 7.9. Silhouettes such as these underscored the irregularity of minority-majority congressional districts crafted for the post-1990 remap.

As the headbands and moose antlers in figure 7.9 attest, redistricting officials, aided by consultants experienced in "carto-surgery," cleverly constructed favorable constituencies that not only met the equal-population mandate and the then-important minority-majority goal but also satisfied the traditional contiguity constraint, which allows outliers only for true islands, surrounded by water.[36] Illinois crafted its earmuffs district (figure 7.9, right) by connecting sizable but separated concentrations of Latino voters with a thin, often only one-block-thick corridor wrapped around a suitably large intervening concentration of African American voters. This delineation created safe districts for Democrats with Hispanic and African American ties—safe seats that survive today because of arguably partisan motives.[37] Although the courts had rejected the original racial rationale for New York's Bullwinkle district (figure 7.9, left), a less ragged version lasted for more than two decades as a safe seat for a Latina Democrat who represents constituents in parts of Manhattan, Queens, and Brooklyn.[38] Bridges and tunnels are acceptable links when a river bisects an otherwise contiguous district.

Race, clearly, was not the only factor behind the irregular boundaries of the 1990 remap, which also reflected partisan efforts to protect incumbents and undermine opponents.[39] Putting two of the other party's incumbents in the same district, in which one was certain to lose, is an effective (if not brutal) tactic. The cartographic weirdness persists because federal judges, however appalled by odd shapes with pejorative labels, have been reluctant to outlaw partisan gerrymandering, deemed the right of a duly elected legislature—"separation of powers" it's called.[40] Not surprisingly, political boundaries delineated in the post-2010 remap rivaled the complexity of their post-1990 ancestors.

A poster child of the persistent exploitation of networked census data for partisan purposes is Maryland's Third District (figure 7.10), which the *New Republic* called "America's most gerrymandered district."[41] Its shape is entangled figuratively and literally with the shapes of its neighbors, all drawn with the intent of electing Democrats—often a specific Democrat—while minimizing the electability of Republicans, conveniently packed into a largely rural and predictably conservative district that covers the entire Eastern Shore as well as Baltimore's outer, more rural suburbs. Seven of the state's eight congressional districts reliably elect Democrats, while the other is consistently Republican. Figure 7.10 illustrates how thin isthmuses no wider than an eighth of a mile in places create a political critter dubbed the "praying mantis."[42] Figure 7.11, an enlargement of the emaciated neck highlighted in figure 7.10, shows portions of District Three at the top and bottom connected by a strand that separates districts immediately to the east and west by no more than a long city block—sufficient nonetheless to satisfy the contiguity requirement. Expanding the silhouette to include a small part of Chesapeake Bay makes the district's outer boundary less irregular and hides disconnections of patches not linked by roads within the district.

I am not impressed by arguments that assign strange powers to maps that are little more than tools for leveraging the clout of those with the skill and position to exploit them. Although digital maps and block census data enable partisan gerrymandering, the electorate is not powerless, as demonstrated in the 2018 midterm elections, when numerous districts that had preferred Trump to Clinton in 2016 contributed to the Democrats' takeover of the House. Until the courts or

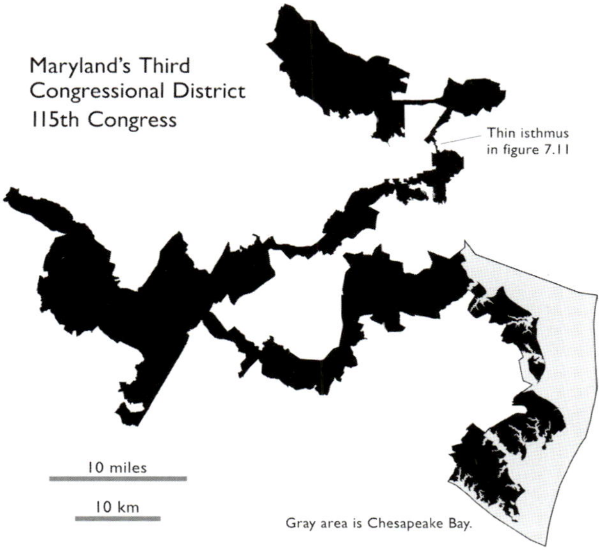

Figure 7.10. Maryland's Third Congressional District, based on 2010 Census data and TIGER/Line shapefiles. Some silhouettes of District Three include Chesapeake Bay; others show only the district's land area.

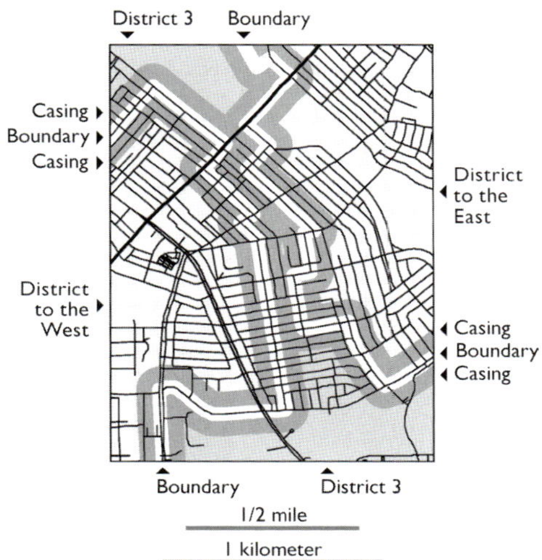

Figure 7.11. Greatly enlarged excerpt for the boundary segment labeled a thin isthmus in figure 7.10. Adapted from a Census Bureau map that highlights district boundaries with a pair of gray casings, one inside and the other outside the district. A third of the distance up from the bottom of the map, the interior casings merge because the narrow strip of land joining portions of District Three at the top and bottom of this excerpt is no wider than a long city block.

Congress rein in partisan gerrymandering, networked street data will help cunning politicians fabricate questionable majorities.

Lest this seem an awkward note on which to conclude both the book and the chapter, let me refer to the original goal, outlined in the preface, of exploring the diverse connections between maps and networks. As strategies for navigating routeways, seeing patterns, or planning projects, maps depend on networks of measurements, observations, and other data to provide the content portrayed by cartographic symbols. And many networks, especially the bigger ones that are part of the regional, national or international infrastructure, broadly defined, depend (or depended) on maps for their design, planning, construction, maintenance, and continued operation. The foregoing stories reinforce the overriding hypothesis that the history of cartography and the history of networks are deeply and meaningfully intertwined.

APPENDIX A
IMAGE CREDITS

Chapter 1

Figure 1.3: NOAA Photo Library, Historic Coast & Geodetic Survey (C&GS) Collection, Plate 30, Coast Survey Report 1865.

Figure 1.5, top: William Gillespie, *Treatise on Land-Surveying*, 307.

Figure 1.5, bottom: Gillespie, *Treatise on Land-Surveying*, 306.

Figure 1.6: Gillespie, *Treatise on Land-Surveying*, 213.

Figure 1.7: Gillespie, *Treatise on Land-Surveying*, 265.

Figure 1.8: NOAA Photo Library, Historic Coast & Geodetic Survey (C&GS) Collection, image theb1647.

Figure 1.9: redrawn from Charles Anthony Schott, *The Eastern Oblique Arc of the United States*, US Coast and Geodetic Survey Special Publication No. 7 (Washington, DC, 1902), map *A*.

Figure 1.10: Portion of Schott, *Eastern Oblique Arc*, map *A*.

Figure 1.11: O[rlando] P[latt] Sutherland, *Triangulation in Massachusetts*, US Coast and Geodetic Survey Special Publication No. 76 (Washington, DC, 1922), map 19.

Figure 1.12: high-resolution version of image posted online at https://s3.amazonaws.com/gs-geo-images/db5b0777-9ae9-4e98-96dc-acdf76d3b6a8_1.jpg by Richard Garland (aka Papa Bear).

Chapter 2

Figure 2.3: John F. Hayford, *A Text-Book of Geodetic Astronomy* (New York: John Wiley & Sons, 1898), fig. 10 (no page, back of book).

Figure 2.4: Hayford, *Text-Book*, fig. 13 (no page, back of book).

Figure 2.5: Hayford, *Text-Book*, figure adapted from figs. 30 and 31 (no page, back of book).

Figure 2.6: redrawn from progress sketch no. 3 in *Report of the Superintendent of the United States Coast Survey Showing the Progress of the Work for the Fiscal Year Ending with June 1876* (Washington, DC, 1878).

Figure 2.7: excerpt from progress sketch no. 3 in *Report of the Superintendent of the United States Coast Survey June 1876.*

Figure 2.8: redrawn from "Telegraphic Longitude Net of the United States and Its Connection with That of Europe," *Report of the Superintendent of the United States Coast and Geodetic Survey Showing the Progress of the Work During the Fiscal Year Ending with June 1897* (Washington, DC, 1898), map follows p. 261.

Figure 2.9: redrawn from the map *Geodetic Control for the 1901 Adjustment*, figure 1 in Maralyn L. Vorhauer, "Horizontal Datums: Knowing Where in the World You Are," National Geodetic Survey, 2007, https://www.ngs.noaa.gov/web/about_ngs/history/datums.gel.pdf.

Figure 2.10: excerpts from US Geological Survey, 1984, Umbrella Creek, Washington, 7.5-minute quadrangle map (provisional).

Figure 2.11: redrawn from image of the "Expandable 24-Slot" satellite configuration provided online by the GPS Directorate at https://www.gps.gov/systems/gps/space/.

Chapter 3

Figure 3.1: Edwin R. Stuart, *Topographical Drawing* (New York: McGraw-Hill Book Co., 1917), 68.

Figure 3.2: excerpt from US Geological Survey, 1900, Oswego, New York, 15-minute quadrangle map.

Figure 3.4: *Topographic Instructions of the United States Geological Survey* (Washington, DC, 1913), 208. The page also includes symbols for fords (3), ferries (2), and a dam (1).

Figure 3.5: excerpts from US Geological Survey, 1898, Syracuse, New York, 15-minute quadrangle map.

Figure 3.6: waterway delineations from War Department, Corps of Engineers, 1903, Oswego Harbor, New York, 1:8,000 navigation chart.

Figure 3.7: excerpt from US Geological Survey, 1900, *Oswego Special Map*, New York, 15-minute quadrangle map.

Figure 3.8: *Topographic Instructions* (Washington, DC, 1913), 207.

Figure 3.9: maps redrawn from War Department, Corps of Engineers, 1903, Oswego Harbor, New York, 1:8,000 navigation chart.

Figure 3.10: excerpt from US Geological Survey, 1965, Rochester, New York, 1:250,000 topographic map.

Figure 3.11, upper left: excerpt from US Geological Survey, 1957, Syracuse East, New York, 7.5-minute quadrangle map.

Figure 3.11, upper center: excerpt from US Geological Survey, 1978, Syracuse East, New York, 7.5-minute quadrangle map, photo revised.

Figure 3.11, upper right: excerpt from New York State Department of Transportation, 1990, Syracuse East, New York, 7.5-minute planimetric map.

Figure 3.11, lower: excerpt from US Geological Survey, 1957, Manlius, New York, 7.5-minute quadrangle map.

Figure 3.12: US National Resources Committee, "Transportation: Railroads, Bus Lines, Subways, Pipe Lines, Etc.," *Suggested Symbols for Plans, Maps, and Charts* (Washington, DC, June 1938), plate F-2.

Figure 3.13: *Official Guide of the Railways and Steam Navigation Lines of the United States*, June 1957, 168.

Figures 3.14 and 3.15: hypothetical place-names are from the first words of random phrases created by the online Lorem Ipsum generator at www.lipsum.com.

Chapter 4

Figure 4.1: various sources include "Map of the Canal System of the State of New York, to accompany Supplement to the Annual Report of the State Engineer and Surveyor" (1905); and "Map Showing the Electric Power Lines in the State of New York, Issued by the Niagara, Lockport, and Ontario Power Company" (Buffalo, NY: Matthews-Northrup Works, ca. 1912).

Figure 4.2: redrawn (with two excerpts) from *Laws of the State of New York, in Relation to the Erie and Champlain Canals* (Albany, NY, 1825), tipped in opp. p. 38.

Figure 4.3: *Laws of the State of New York,* tipped in opp. p. 44.

Figure 4.4, top: excerpt from US Geological Survey, 1902, Clyde, New York, 15-minute quadrangle map.

Figure 4.4, bottom: excerpt from US Geological Survey, 1953, Clyde, New York, 15-minute quadrangle map.

Figure 4.5: redrawn from portion of Parks & Trails New York's interactive trail map *Cycle the Erie Canal*, online at https://www.ptny.org/bike-canal/map/.

Figure 4.6: excerpts from figs. 84 and 86, in William G. Raymond, *Elements of Railroad Engineering* (New York: John Wiley and Sons, 1908), 266–67.

Figure 4.7: Thomas M. Cleemann, *The Railroad Engineer's Practice* (New York: D. Van Nostrand Co., 1892), between pp. 20 and 21.

Figure 4.8: excerpts from Raymond, *Elements of Railroad Surveying*, plate XIV, a foldout inside the back cover.

Figure 4.9: William Beahan, *Field Practice of Railway Location* (New York: Engineering News, 1904), excerpt from plate 2, tipped in after p. 56.

Figure 4.10: William T. Pryor, "The Role of Aerial Surveys in Highway Engineering" (paper presented at the Ninth Congress, International Society for Photogrammetry, London, England, September 5–17, 1960), 12.

Figure 4.11: grid and tick mark facsimiles from US Geological Survey, 1947, Boonville, New York, 7.5-minute quadrangle map. SPC zone boundaries from Hugh C. Mitchell and Lansing G. Simmons, *The State Coordinate Systems (A Manual for Surveyors)*, US Coast and Geodetic Survey Publication no. 235 (Washington, DC, 1945), fig. 2.

Figure 4.12: excerpt from Matthews-Northrup Works, "Map Showing the Electric Power Lines in the State of New York Issued by the Niagara, Lockport, and Ontario Power Company," Buffalo, New York, ca. 1912.

Figure 4.13: Carl Frederick Meyer, *Route Surveying* (Scranton, PA: International Textbook Co., 1949), fig. 11-1, 270.

Chapter 5

Figure 5.1: Wilhelm Trabert, *Meteorologie und Klimatologie* (Leipzig: F. Deuticke, 1905), 65.

Figure 5.2: map compiled by the author using HamsterMap.com to plot spherical coordinates on a Google Maps base framed on a Web Mercator projection.

Figure 5.3: map compiled from lists of meteorological stations and observers, organized by state, in *Smithsonian Report*, 1849, 27–31, and 1860, 87–100.

Figure 5.4: Smithsonian Institution, *Map Showing the Stations Where Meteorological Observations Are Made in the United States under the Direction of the Smithsonian Institution [in] 1851*, scale 1:6,500,000 (Washington, DC, 1851). Excerpt from a scanned image provided by the Geography and Map Division, Library of Congress.

Figure 5.5: compiled from lists in *Smithsonian Reports*, 1858, 73; 1859, 95; and 1860, 101, and from network delineations in Chas. B. Barr, *Telegraph Stations in the United States, the Canadas, and Nova Scotia* (Pittsburgh, PA: Wegner & Buechner, 1853).

Figure 5.6: scan provided by the Geography and Map Division, Library of Congress. Catalog information online at https://lccn.loc.gov/98686206.

Figure 5.7: compiled from table VI, "Showing telegraphic circuits used for weather reports," *Annual Report of the Chief Signal-Officer to the Secretary of War for the Year 1871*, 57.

Figure 5.8: W. B. Hazen, *History of the Signal Service, with Catalogue of Publications, Instruments, and Stations [Prepared under the direction of Brig. & Bvt. Maj. Gen'l W. B. Hazen, Chief Signal Officer of the Army]* (Washington City, 1884), 23.

Figure 5.9: Robert DeCourcy Ward, "The Newspaper Weather Maps of the United States," *American Meteorological Journal* 11 (July 1894): 96–107; unnumbered page between 100 and 101.

Figure 5.10: compiled from illustrations in Ward, "The Newspaper Weather Maps of the United States," unnumbered pages between 100 and 101.

Figure 5.11: Frank H. Bigelow, *Storms, Storm Tracks, and Weather Forecasting*, US Weather Bureau Bulletin no. 20 (1897), chart 10, on 40.

Figure 5.12, top: compiled from *Operations of the Weather Bureau, April 1, 1967* (Silver Spring, MD: US Dept. of Commerce, Environmental Services Administration, Weather Bureau, 1967), 21.

Figure 5.12, bottom: compiled from Committee on the Assessment of the National Weather Service's Modernization Program, Board on Atmospheric Sciences and Climate, National Research Council, *The National Weather Service Modernization and Associated Restructuring: A Retrospective Assessment* (Washington, DC: National Academies Press, 2012), fig. 4.10b on p. 60; and National Weather Service, "List of National Weather Service Weather Forecast Offices," confirmed as current in May 2018 and provided online at National Weather Service (Sept. 19, 2012). "National Weather Service Organization." weather.gov. Retrieved Feb. 22, 2013.

Figure 5.13: compiled from NEXRAD Panel, National Weather Service Modernization Committee, Commission on Engineering and Technical Systems, National Research Council, *Toward a New National Weather Service: Assessment of NEXRAD Coverage and Associated Weather Services* (Washington, DC: National Academy Press, 1995), figs. 2.1a and 2.1b, on p. 9.

Figure 5.14, left: compiled from National Weather Service lists and map of forecast offices and coverage areas at National Weather Service, Offices and Centers, "NWS Weather Offices by Region," accessed May 11,2018, http://www.nws.noaa.gov/organization.php; List of Weather Service Offices, accessed May 11, 2018, https://www.weather.gov/media/tg/wsr88d-radar-list.pdf; and http://www.nws.noaa.gov/organization.php.

Figure 5.14, right: compiled from List of NEXRAD sites, 11/29/2016, accessed May 11, 2018, "WSR-88D Radar List, 159 sites," https://www.weather.gov/media/tg/wsr88d-radar-list.pdf.

Figure 5.15: compiled from Committee on the Assessment of the National Weather Service's Modernization Program, Board on Atmospheric Sciences and Climate, National Research Council, *The National Weather Service Modernization and Associated Restructuring: A Retrospective Assessment* (Washington, DC: National Academies Press, 2012), fig. 4.1 on p. 48; and individual state-specific ASOS maps provided online at Federal Aviation Administration, Surface Weather Observation Stations, ASOS/AWOS, accessed May 10, 2018, https://www.faa.gov/air_traffic/weather/asos/.

Chapter 6

Figure 6.2, left: Frank Z. Hutton Jr. and C. Erwin Rice, Soil Survey of Onondaga County, New York (Washington, DC: US Department of Agriculture, Soil Conservation Service, 1977), excerpt from sheet 4.

Figure 6.2, right: US Geological Survey, 1980, *Land Use and Land Cover* (1975–77), Sherbrooke, Canada; United States, 1:250,000.

Figure 6.6: map created by author from screen captures. The Dinosaur National Monument Trails dataset in ArcGIS® Online (https://www.arcgis.com/index.html) is based on USGS National Map data.

Figure 6.8: parts (B) and (C) in example 5, Gavin H. Shaw, CREATE software, *A Supervisor's Manual: Supervisory Procedures for the Creation or Extension of a GBF/DIME System* (Washington, DC: Bureau of the Census, 1976), 30.

Figure 6.9: George Farnsworth, *Census Use Study: DIME, a Geographic Base File Package* (Washington, DC: Bureau of the Census, 1972), 37.

Figure 6.10: adapted from fig. 11 in Marvin S. White Jr., "A Geometrical Model for Error Detection and Correction," in *Proceedings of the International Symposium on Computer-Assisted Cartography, Auto-Carto III*, January 16–20, 1978 (Falls Church, VA: American Congress on Surveying and Mapping, 1979), 452.

Figure 6.12, left: the Waze® iPhone® app (version 4.3.3) in navigation mode, Wikimedia Commons, November 29, 2017, https://commons.wikimedia.org/w/index.php?title=File:Waze_4.3_iPhone_app.jpg&oldid=270205093.

Figure 6.12, right: screenshot by the author.

Figure 6.13: Alan Polivka, Bill Moore Ede, and Joe Drapa, *North American Joint Positive Train Control Project*, Report DOT/FRA/ORD-09/04 (Washington, DC: Federal Railroad Administration, 2009), 7.

Chapter 7

Figure 7.1: elements reconfigured slightly from DARPA image in Waldrop, "DARPA and the Internet Revolution," 78.

Figure 7.2: F. Heart, A. McKenzie, J. McQuillian, and D. Walden, *ARPANET Completion Report*, BBN Technical Report 4799 (Cambridge, MA: Bolt Beranek and Newman, January 1978), 203.

Figure 7.3: Heart et al., *ARPANET Completion Report*, BBN Technical Report, 137.

Figure 7.4: adapted from color maps posted at National Science Foundation, "Sequential Maps of Internet Growth: Four Maps of Internet Growth in the United States: 1969 ARPANET, 1977 ARPANET, 1988 NSFNET Backbone, 1995 NSFNET T3 Backbone," accessed October 4, 2018, https://www.nsf.gov/news/special_reports/nsf-net/60s_to_90s_map.pdf.

Figure 7.5: adapted from a color map posted at National Science Foundation, "When the Country Called: How a Team of Academic Experts Contributed to the President's Cyberspace Review," News Release 09-115, May 29, 2009.

Figure 7.6: adapted from a color map posted at Defense Advanced Research Projects Agency, accessed October 9, 2018, https://www.darpa.mil/ddm_gallery/sigma_map.jpg.

Figure 7.7: compiled from Federal Election Commission, *Election Results for the US President, the US Senate, and the US House of Representatives*, December 2017, accessed October 11, 2018, https://transition.fec.gov/pubrec/fe2016/federalelections2016.pdf.

Figure 7.8: compiled from *Daily Kos* staff, "Presidential Results by Election District: Results for Districts Used in 2016," accessed October 16, 2018, https://www.dailykos.com/stories/2018/2/21/1742660/-The-ultimate-Daily-Kos-Elections-guide-to-all-of-our-data-sets, February 21, 2018.

Figure 7.9: New York's Twelfth District and Illinois's Fourth District, compiled from US Bureau of the Census, *Congressional District Atlas*, 103rd Congress of the United States (Washington, DC, 1993).

Figure 7.10: adapted from US Bureau of the Census, *115th Congress of the United States, Maryland—Congressional District 3*, accessed October 25, 2018, https://www2.census.gov/geo/maps/cong_dist/cd115/cd_based/ST24/CD115_MD03.pdf.

Figure 7.11: adapted from US Bureau of the Census, *115th Congress of the United States, Maryland – Congressional District 3* map.

APPENDIX B
ACRONYMS

ACG: address coding guide
ARPA: Advanced Research Projects Agency (predecessor of DARPA)
ARPANET: Advanced Research Projects Agency Network
ASOS: Automated Surface Observing System
BITNET: Because It's Time Network
C&GS: Coast and Geodetic Survey
CONUS: conterminous (contiguous) United States
CSNET: Computer Science Network
CSX: corporation formed by the merger of the Chessie System and the Seaboard Coast Line railways
CSXT: CSX Transportation
DARPA: Defense Advanced Research Projects Agency
DEM: digital elevation model
DIME: dual independent map encoding
DOCB: Department of Commerce Boulder Laboratories
DOT: Department of Transportation
FAA: Federal Aviation Administration
FRA: Federal Railroad Administration
GBF: geographic base file
GDA2020: Geocentric Datum of Australia 2020
GDA94: Geocentric Datum of Australia 1994
GIS: geographic information science (academic subfield) in some contexts, geographic information system (software + data) in others
GLONASS: Globalnaya Navigazionnaya Sputnikovaya Sistema (Russia's Global Navigation Satellite System)

GPS: Global Positioning System
GRS80: Geodetic Reference System of 1980
IBM : intercontinental ballistic missile
IERRSS: International Earth Rotation and Reference Systems Service
IMP: interface message processor
ITIS: International Terrestrial Information Service
MILNET: Military Network
NAD27: North American Datum of 1927
NED: National Elevation Dataset
NEXRAD: Next Generation Weather Radar
NGS: National Geodetic Survey
NGVD29: National Geodetic Vertical Datum of 1929
NOAA: National Oceanic and Atmospheric Administration
NORSAR: Norwegian Seismic Array
NSFNET: National Science Foundation Network
NWS: National Weather Service
O&W: New York, Ontario, and Western Railway
OCR: optical character recognition
PTC: positive train control
SLR: satellite laser ranging
SPCS: state plane coordinate system
TIGER: topologically integrated geographic encoding and referencing
TIN: triangulated irregular network
TIP: terminal interface processor
UAV: unmanned aerial vehicle
USGS: US Geological Survey
VLBI: very long baseline interferometry
WAAS: Wide Area Augmentation System
WFO: Weather Forecast Office
WGS84: World Geodetic System of 1984
WSFO: Weather Service Forecast Office
WSO: Weather Service Office
WSR-88D: Weather Surveillance Radar, 1988, Doppler

NOTES

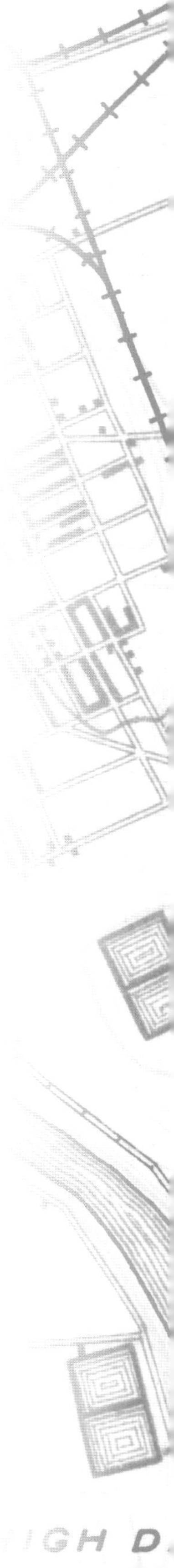

Chapter 1

1. Arthur H. Robinson and Barbara Bartz Petchenik, *The Nature of Maps: Essays toward Understanding Maps and Mapping* (Chicago: University of Chicago Press, 1976), 14–17.
2. Mark Monmonier, *Adventures in Academic Cartography: A Memoir* (Syracuse, NY: Bar Scale Press, 2014), 25–44.
3. Peter Haggett and Richard J. Chorley, *Network Analysis in Geography* (London: Edward Arnold, 1969), 8, 31, and 47.
4. Although geospatial technology uses horizontal distance and vertical distance when addressing concepts of horizontal and vertical accuracy, planimetric distance focuses on the historically important practice of drafting and reproducing most maps as networks that exist in a plane.
5. Peter Collier, "Plane Table," in *The History of Cartography*, vol. 6: *Cartography in the Twentieth Century*, ed. Mark Monmonier (Chicago: University of Chicago Press, 2015) [hereafter *HC6*], 1155–57.
6. W[illiam] M[itchell] Gillespie, *Treatise on Land-Surveying: Comprising the Theory Developed from Five Elementary Principles; and the Practice with the Chain Alone, the Compass, the Transit, the Theodolite, the Plane Table, &c.* (New York: D. Appleton & Co., 1855), quotation on 307.
7. For example, the Irregular Triangle Calculator at handymath.com presents six boxes labeled Side *A*, Side *B*, Side *C*, Angle *A* (opposite Side *A*), Angle *B* (opposite Side *B*), and Angle *C* (opposite side *C*). Fill in any three boxes, and the calculator will return results in the other three.
8. Here's the underlying trigonometry for the first two calculated lengths: if the lines of sight are accurately plotted, the line to the distant intersection point (to be calculated) and half the baseline (12.5 feet) are the opposite and adjacent sides of a right triangle with a hypotenuse of 500 ft. Dividing 12.5 by 500 yields a cosine of 0.025 for the angle opposite the length to be calculated, and finding the arc cosine yields an angle of 88.57°, which has a tangent of 39.9875. My calculator, no doubt subject to a small amount of rounding error, returned several more decimal places, which I have rounded down. Because the tangent is

the ratio of the length of the opposite side to the length of the adjacent side, multiplying the tangent by 12.5 yields 499.84 ft. for the distance between the intersection point and the baseline. By contrast, if the lines of sight are underestimated by a tenth of a degree, the resulting angle 89.47° yields a tangent of 37.3771 and a distance of 467.21 ft. from the intersection to the baseline.

9. Gillespie, *Treatise on Land-Surveying*, 263–64. Consider this example, using triangulations based on my hypothetical 25-ft. baseline (acute angle = 2.9°), which is highly sensitive to angular discrepancies of a quarter degree, a half degree, and a full degree, which yield positional discrepancies of 74.3, 129.4, and 205.6 ft., respectively. By contrast, a 100-ft. baseline (acute angle = 11.5°) reduces the discrepancies to 20.9, 40.1, and 74.4 ft., respectively, while a 500-ft, baseline, which makes this an equilateral triangle, yields corresponding errors of only 4.3, 8.6, and 16.9 ft.
10. For a more current reference, see Lajos Fialovszky and Gyula Holéczy, eds., *Surveying Instruments and Their Operational Principles* (Amsterdam: Elsevier, 1991), esp. 75–194. Also see Gillespie, *Treatise on Land-Surveying*, 211–39.
11. Gillespie, *Treatise on Land-Surveying*, 223. This increased precision relied on short supplementary scales called *verniers*, which afforded more accurate estimates of fractional degrees. In figure 1.6, the letter *V* identifies a vernier that increased the reliability of horizontal angles.
12. *Report of the Superintendent of the Coast Survey, Showing the Progress of the Survey during the Year 1854* (Washington, DC: A. O. P. Nicholson, Public Printer, 1855), 78.
13. Gillespie, *Treatise on Land-Surveying*, 267.
14. US Coast Survey, "Sketch *A* Showing the Primary Triangulation in Section No. 1, from 1844 to 1855" (scale 1/600,000, 1854), NOAA's Historic Coast & Geodetic Survey (C&GS) Collection image 05853. This map allowed me to add the bar scale, missing from Gillespie's drawing.
15. An online Irregular Triangle Calculator yielded the 13.13-mi. result after I entered 10.77 mi. as the length of the baseline, 54.11° as the angle opposite the baseline, and 81.00° as the angle at South Base. These angles in decimal degrees were rounded from the more precise measurements 54°06'45.246" and 81°00'16.660" reported for triangle no. 65 in the table "Resulting Angles and Distances of the Primary Triangulation, etc." in *Report of the Superintendent of the United States Coast Survey Showing the Progress of the Survey During the Year 1865* (Washington, DC, 1867), 200. The length of the baseline in miles was converted (and rounded) from the estimated length of 17,326.3763 m reported in *Report of the Superintendent 1865*, 189.
16. US Coast Survey, "Sketch *A* Showing the Primary Triangulation in Section No. 1, from 1844 to 1855." http://www.photolib.noaa.gov/bigs/cgs05853.jpg.
17. "Appendix No. 21. Results of the Primary Triangulation of the Coast of New England, from the Northeastern Boundary to the Vicinity of New York," 187–203 in *Report of the Superintendent of the Survey of the Coast, Showing the Progress of the Survey During the Year 1865* (Washington, DC, 1867), quotation on 187. The appendixes of the report were published collectively as H.R. Exec. Doc. No. 75, 39th Cong., 1st Sess. (1865–66).

18. *Report of the Superintendent of the Survey of the Coast, Showing the Progress of the Work During the Year Ending November 1844* (Washington, DC: Gales and Seaton, 1845), S. Doc. No. 16, 28th Cong., 2nd sess. (1844), esp. 6–9.
19. Albert E. Theberge, *The Coast Survey: 1807-1867* [vol. I of the *History of the Commissioned Corps of the National Oceanic and Atmospheric Administration*] (Silver Spring, MD: US Dept. of Commerce, National Oceanic and Atmospheric Administration, NOAA Central Library, 1998), 140–43.
20. Richard D. Cutts, "Field-Work of the Triangulation" (US Coast Survey, 1877), quotation on 40. On the title page, directly below the byline, was the note, "Reprinted, with additions, from the Coast Survey Report of 1868."
21. "Appendix No. 21," 187–89; and Lewis V. Hudson, *Weights and Measures Standards of the United States: A Brief History*, National Bureau of Standards Special Publication 447 (Washington, DC, 1963, 1976), esp. 6. The Coast Survey report put the length of the scale, manufactured by Troughton of London, at 84 in., in contrast to Hudson, who made it 82 in.
22. F. R. Hassler, "Description of the Apparatus for Measuring Base Lines," pages 273–86 in his "Papers on Various Subjects Connected with the Survey of the Coast of the United States," *Transactions of the American Philosophical Society* new series, 2 (1825): 232–420. Also see E. B. Hunt, "Appendix No. 35. Description of the United States Coast Survey Apparatus for Measuring Base Lines," *Report of the Superintendent of the Survey of the Coast Survey for 1854*, 103–8.
23. *Report of the Superintendent of the United States Coast Survey Showing the Progress of the Survey During the Year 1865* (Washington, DC, 1867), 189.
24. Cutts, "Field-Work of the Triangulation," 3.
25. *Arc* has a more general meaning in geometry, where it refers to part of a circular curve. In geospatial technology, *arc* refers to a curve or line defined by a series of x,y coordinates, which explains its role as the lead syllable in the names of Esri's software products, such as ArcGIS® and ArcMap®.
26. Cutts, 9–11.
27. For a more detailed discussion, see J. Clendinning, "Mathematical Theory of the Transit Theodolite," *Empire Survey Review* 6 (1941): 194–205, 462–74; W. Schofield and M. Breach, *Engineering Surveying* (Abingdon, Oxon, UK: Spon Press, 2011), 160–88; and Paul Kunkel, "Theodolite Error," January 26, 2012, http://whistleralley.com/surveying/theoerror/.
28. Edward L. Ingram, *Geodetic Surveying and the Adjustment of Observations (Method of Least Squares)* (New York: McGraw-Hill Book Co., 1911), 81–102; quotation on 88.
29. For a critique of Ingram's approach, which might not be the best way to accommodate adjustments of angles and lengths based on different measurement technologies, see J. E. Lilly, "Least Squares Adjustment of Dissimilar Quantities," *Empire Survey Review* 16 (1961): 120–28.
30. Charles A. Schott, *The Eastern Oblique Arc of the United States*, US Coast and Geodetic Survey Special Publication No. 7 (Washington, DC, 1902). For an account of Schott's accomplishments, see Cleveland Abbe, *Biographical Memoir of Charles Anthony*

Schott, 1826–1901, National Academy of Sciences Biographical Memoirs, part of vol. 3 (Washington, DC, 1915).

31. See, for example, W. Norman Thomas, *Surveying* (London: Edward Arnold, 1920), 125–46.
32. For an example of the calculation of a network's average closing error, expressed as an angle, see William Bowie, *Primary Triangulation of the One Hundredth and Fourth Meridian, and on the Thirty-Ninth Parallel in Colorado, Utah, and Nevada*, US Coast and Geodetic Survey Special Publication No. 19 (Washington, DC, 1914), 56–63.
33. Schott, *Eastern Oblique Arc*, 235.
34. Schott called this three-dimensional figure an "osculating spheroid." An oblique arc addresses the spheroid's circular cross sections parallel to the plane of the equator as well as its elliptical cross sections anchored to the planet's axis of rotation. See Schott, 20.
35. My black-and-white rendering adjusts for a small but noticeable printing error, namely, the slight shift of the orange tint toward the bottom of the print, leaving a white gap along the upper edge of the orange area and overlap with the blue area along the lower (north–south trending) edge. Failure to precisely align impressions made with different inks is known as *misregistration*.
36. Schott, 42. The small map also shows a quadrilateral described as "an integral part" of the "direct connection with the base" but extending eastward beyond the orange tint on Map *A*.
37. Schott, 29, 47, 58, 76, and 93. As with the Massachusetts Base, the orange areas on Map *A* do not include all the quadrilaterals on the small maps.
38. O[rlando] P[latt] Sutherland, *Triangulation in Massachusetts*, US Coast and Geodetic Survey Special Publication No. 76 (Washington, DC, 1922). The maps follow p. 283.
39. Sutherland, *Triangulation*, 2 and 3.
40. Deborah Jean Warner, "Electronic Distance Measurement," *HC6*, 355–56.
41. Papa-Bear-NYC, "The Eastern Oblique Arc Crosses Massachusetts: A Long Weekend Searching for Old Triangulation Stations," April 25, 2008, http://forums.groundspeak.com/GC/index.php?showtopic=192058. For general information on the hobby, see Benchmark Hunting, https://www.geocaching.com/mark/.
42. Data for the station Massachusetts North Base are online at National Geodetic Survey (NGS), Survey Marks and Datasheets, https://www.ngs.noaa.gov/datasheets/ under the station's name or its PID (permanent identifier) MY3577.
43. National Geodetic Survey, "Massachusetts South Base 1844," Survey Marks and Datasheets, PID LW2602.
44. Papa-Bear-NYC, "The Epping Base Net of the Eastern Oblique Arc: Fifth of a Series," November 10, 2008, http://forums.groundspeak.com/GC/index.php?showtopic=207678.
45. Burni Andres, "150th Anniversary of the Epping Baseline," Cherryfield Narraguagus Historical Society, stories (July 1, 2007), http://www.cherryfieldhistorical.com/2007/07/150th-anniversary-of-the-epping-baseline/. The line was not level, but at each step the deviation from the horizontal was measured and used later to correct for the inclination.
46. National Geodetic Survey, "Epping Base East" and "Epping Base West," Survey Marks and Datasheets, PD1145 and PD11, respectively.

47. Papa-Bear-NYC, "The Epping Base Net." The length is confirmed by "Appendix 21: Results of the Primary Triangulation."
48. Andrew Kekacs, "Epping's Plains Site of Ambitious 1850s Survey," *Bangor Daily News*, November 23–24, 1991, Maine Weekend section, 12.
49. The policy was rescinded on May 2, 2000, when the White House bowed to the consumer electronics industry, which was eager to sell reliable personal and in-vehicle GPS receivers. Mark Monmonier, *Spying with Maps: Surveillance Technologies and the Future of Privacy* (Chicago: University of Chicago Press, 2002), 14–15.
50. For information on the technical work-around, known as DGPS (differential GPS), see Horst Hecht, "Global Positioning Systems in Hydrographic Mapping," *HC6*, 632–33; and Jacques Beser and Bradford W. Parkinson, "The Application of NAVSTAR Differential GPS in the Civilian Community," *Navigation* 29 (1982): 107–36.
51. The calculations, along with various adjustments, are discussed in Harold E. Nelson and Joseph F. Dracup, "A Tale of Two Eras in American Geodesy," *Surveying and Land Information Systems* 60 (2000): 163–76, esp. 173–76. The baseline was measured again, in 1992, by an NGS survey party.
52. National Geodetic Survey, "Epping Base East" and "Epping Base West."
53. For a link and description of the application INVERSE, see NGS Geodetic Toolkit, https://www.ngs.noaa.gov/TOOLS/. My result differs from the calculation from the 1992 survey, reported by Nelson and Dracup, by only 0.0002 m.
54. Günter Seeber, *Satellite Geodesy*, 2nd ed. (Berlin: Walter de Gruyter, 2003), 140, 162.

Chapter 2

1. For a concise and now classic overview of historical trends in modes of mapping, see Matthew H. Edney, "Cartography without 'Progress': Reinterpreting the Nature and Historical Development of Mapmaking," *Cartographica* 30, nos. 2–3 (1993): 54–68. For a much updated, more comprehensive treatment, see his *Cartography: The Ideal and Its History* (Chicago: University of Chicago Press, 2019).
2. Helen M. Wallis and Arthur H. Robinson, *Cartographical Innovations: An International Handbook of Mapping Terms to 1900* (Tring, Herts, UK: Map Collector Publications, 1972), 172–74.
3. The position of Polaris in the sky is continuous and predictable, and can be looked up in an ephemeris. See James A. Van Allen, "Basic Principles of Celestial Navigation," *American Journal of Physics* 72 (2004): 1418–24.
4. "Celestial Navigation: Navy Resurrects Ancient Craft to Thwart Hackers," *Progressive Digital Media Defense* (Incl. Air Force, Army, Navy and Homeland Security) *News*, February 29, 2016; David Dickinson, "Navy Resumes Celestial Navigation Course," *Sky and Telescope* 132 (July 2016): 12; and Andrea Peterson, "U.S. Midshipmen Are Learning the Ancient Art of Sailing by the Stars," *Washington Post*, February 18, 2016.

5. D. H. Boteler, "The Super Storms of August/September 1859 and Their Effects on the Telegraph System," *Advances in Space Research* 38 (2006) 159–72; and Richard A. Lovett, "The Day the Sun Exploded," *Analog Science Fiction & Fact* 132 (November 2012): 21–27.
6. For "line of action of gravity," see John F. Hayford, *A Text-Book of Geodetic Astronomy* (New York: John Wiley & Sons, 1898), 14.
7. Alexander Ross Clarke, *Geodesy* (Oxford: Clarendon Press, 1880), 287–301; and Hayford, Text-Book, 15–16.
8. For a description of the use of the zenith telescope and subsequent calculations, see Hayford, *Text-Book*, 150–96. For observing lists, see 156–60.
9. For examples of tables listing some events for Jupiter's satellites in decimal seconds, see US Bureau of Equipment, *American Ephemeris and Nautical Almanac for the Year 1898* (Washington, DC, 1895), 463–87.
10. Hayford, *Text-Book*, 19–23.
11. Hayford, 75–82.
12. Hayford, 86–87.
13. Hayford, 144.
14. Hayford, 144.
15. Hayford, 167.
16. Dava Sobel, *Longitude: The True Story of a Lone Genius Who Solved the Greatest Scientific Problem of His Time* (New York: Walker and Co., 1995).
17. Hayford, *Text-Book*, 253.
18. Richard Stachurski, *Longitude by Wire: Finding North America* (Columbia, SC: University of South Carolina Press, 2009), 88–124.
19. Hayford, *Text-Book*, 96–98.
20. Standard astronomers' telescopes typically included not one but seven or more wires. Recording the start and end of a star's crossing of each wire afforded a more accurate determination than an observation based on a single wire.
21. Hayford, *Text-Book*, 105–7.
22. Hayford, 242–43; and Stachurski, *Longitude by Wire*, 116–19.
23. Hayford, *Text-Book*, 243–44.
24. Hayford, 249–51.
25. Hayford, 251.
26. "Chart showing longitude-stations and connections determined by the US Coast Survey by means of the electric telegraph between 1846 and 1877, incl.," progress sketch no. 3 in *Report of the Superintendent of the United States Coast Survey Showing the Progress of the Work for the Fiscal Year Ending with June 1876* (Washington, DC, 1878).
27. Stachurski, *Longitude by Wire*, 91–93; and *Report from the Superintendent of the Coast Survey Showing the Progress of the Survey During the Year Ending November 1846*, S. Doc. no. 3, 29th Cong., 2nd sess., 2–3, 72–74.
28. Charles A. Schott, *The Eastern Oblique Arc of the United States*, US Coast and Geodetic Survey Special Publication No. 7 (Washington, DC, 1902), 21.

29. William Frank Zornow, "Jeptha H. Wade in California: Beginning the Transcontinental Telegraph," *California Historical Society Quarterly* 29 (1950): 345–56.
30. *Report of the Superintendent of the United States Coast Survey Showing the Progress of the Survey during the Year 1861* (Washington, DC, 1862), 10.
31. *Report of the Superintendent of the United States Coast Survey Showing the Progress of the Survey during the Year 1869* (Washington, DC, 1872), 48-49; and Oscar Lewis, *George Davidson: Pioneer West Coast Scientist* (Berkeley: University of California Press, 1954), 67–75. For additional information on Davidson, see Charles B. Davenport, "Biographical Memoir of George Davidson, 1825–1911," *Biographical Memoirs*, National Academy of Sciences 18 (1937): 189–217; George W. Dickie, Ralph Harrison, and Samuel B. Christy, "George Davidson: Biographical Sketch," *Proceedings of the California Academy of Sciences*, 4th ser., 4 (1914): 10–13; and Henry R. Wagner, "George Davidson, Geographer of the Northwest Coast of America," *California Historical Society Quarterly* 11 (1932): 299–320.
32. Lewis, *George Davidson*, 35–40.
33. Wagner, "George Davidson," 301.
34. Lewis, *George Davidson*, 40.
35. Benjamin Apthorp Gould, "Appendix No. 6: On the Longitude between America and Europe from Signals through the Atlantic Cable," *Report of the Superintendent of the United States Coast Survey Showing the Progress of the Survey during the Year 1867* (Washington, DC, 1869), 57–133, esp. 60–61.
36. *Report of the Superintendent of the United States Coast Survey Showing the Progress of the Survey during the Year 1854* (Washington, DC, 1855), 11.
37. For a concise chronology of the laying of transatlantic cables in the 1860s, see Bill Glover, "Cable Timeline: Atlantic Cables, 1856-2012," History of the Atlantic Cable & Undersea Communications, last revised October 9, 2016, http://atlantic-cable.com/Cables/CableTimeLine/atlantic.htm.
38. Gould, "Appendix No. 6: Longitude, Transatlantic," 60–61.
39. Gould, 65.
40. Gould, 115.
41. Gould, 63–65.
42. For discussion of three types of exchange used to determine longitude, see Trudy E. Bell, "The Victorian Global Positioning System," *The Bent of Tau Beta Pi 93*, no. 2 (Spring 2001): 14–21.
43. Gould, "Appendix No. 6: Longitude, Transatlantic," 64.
44. Gould, 65.
45. Gould, 113.
46. Gould, 116.
47. Anon., "The French Atlantic Cable: Brest-Duxbury, 1869," History of the Atlantic Cable & Undersea Communications, last revised March 19, 2017, http://atlantic-cable.com/Article/1869French/.

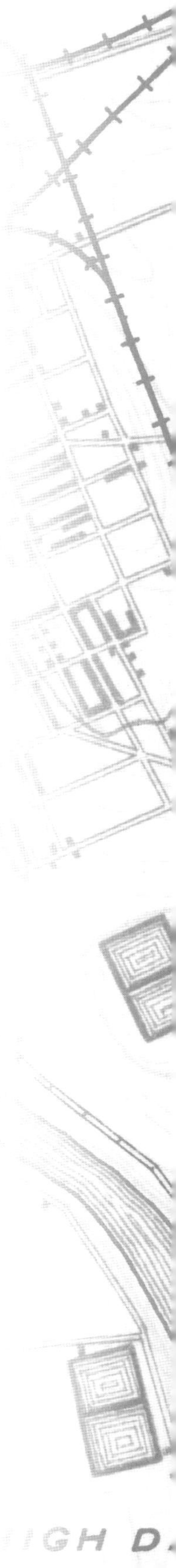

48. J. E. Hilgard, "Appendix No. 18: Transatlantic Longitudes," *Report of the Superintendent of the United States Coast Survey Showing the Progress of the Survey during the Year 1874* (Washington, DC, 1877), 163–242, esp. 163–64.
49. W. J. Brown, "Canada and the Atlantic Cables," *Canada: An Encyclopedia of the Country* (Toronto: Linscott Publishing Co., 1900), 6:536–43; and C. A. Schott, "Appendix No. 2: The Telegraphic Longitude Net of the United States and Its Connection with That of Europe," *Report of the Superintendent of the Coast and Geodetic Survey Showing Progress of the Work during the Fiscal Year Ending with June 1897* (Washington, DC, 1898), 198–261, esp. 201, 203.
50. *Annual Report of the Superintendent of the Coast Survey Showing the Progress of that Work During the Year Ending November 1851* (Washington, DC, 1852), 480.
51. *Report of the Superintendent of the US Coast and Geodetic Survey Showing the Progress of the Work during the Fiscal Year Ending with June 1897* (Washington, DC, 1898), 201.
52. *Report of the Superintendent of the US Coast and Geodetic Survey Showing the Progress of the Work during the Fiscal Year Ending with June 1884* (Washington, DC, 1885), 423, and *Report for 1897*, 254.
53. Schott, "Appendix No. 2: The Telegraphic Longitude Net," 202. Lack of an end-of-century comparison reflects the designation of Lafayette Park as San Francisco's official longitude station.
54. Schott, 201.
55. For workforce statistics for the following fiscal year, see *Report of the Superintendent for 1897*, 8–10. For Schott's annual report for the computing division, see *Report of the Superintendent for 1897*, 80–83. Tide predictions, which were important to hydrographic charting, also relied on an analog (mechanical) computer. See Mark Monmonier, *Coast Lines: How Mapmakers Frame the World and Chart Environmental Change* (Chicago: University of Chicago Press, 2008), 19–21.
56. Schott, "Appendix No. 2: The Telegraphic Longitude Net," 201.
57. Walter Fricke, "Friedrich Wilhelm Bessel (1784–1846)," *Astrophysics and Space Science* 110 (1985): 11–19.
58. C. F. Arden-Close, "A Great British Geodesist: Alexander Ross Clarke (1828–1914)," *Empire Survey Review* 7, no. 49 (1943): 107–16.
59. Arthur R. Hinks, *Maps and Survey*, 5th ed. (Cambridge, UK: Cambridge University Press, 1944), 229–30.
60. "Appendix No. 6: Tables for the Projection of Maps, Based upon a Polyconic Development of the Clarke Spheroid, and Computed from the Equator to the Pole," *Report of the Superintendent for 1884*, 135–321, esp. 135–36.
61. Aaron L. Shalowitz, *Shore and Sea Boundaries, with Special Reference to the Interpretation and Use of Coast and Geodetic Survey Data*, USG&GS publication 10-1 (Washington, DC, 1964), 122–25; and Maralyn L. Vorhauer, "Horizontal Datums: Knowing Where in the World You Are," *National Geodetic Survey*, 2007, https://www.ngs.noaa.gov/web/about_ngs/history/datums.gel.pdf.

62. *Annual Report of the Superintendent, United States Coast and Geodetic Survey to the Secretary of Commerce for the Fiscal Year Ending June 30, 1913* (Washington, DC, 1913), 14–15.
63. George l. Hosmer, *Geodesy, Including Astronomical Observations, Gravity Measurements, and Method of Least Squares* (New York: John Wiley & Sons, 1919), 159–60.
64. Ironically perhaps, in 1924 the International Union of Geodesy and Geophysics (IUGG) adopted the Hayford 1909 ellipsoid for international use. See chap. 2, "Figure of the Earth," in Richard K. Burkard, *Geodesy for the Layman*, US Defense Mapping Agency Technical Report 80-003 (December 1983), https://www.ngs.noaa.gov/PUBS_LIB/Geodesy4Layman/geo4lay.pdf.
65. Shalowitz, *Shore and Sea Boundaries*, 42–52; and E. M. Douglas, "D., Leveling," in *Topographic Instructions of the United States Geological Survey*, eds. William Major Beaman and C. H. Birdseye, *USGS Bulletin* 788D (Washington, DC, 1928), 117–60.
66. Although the change from triangulation, requiring the measurement of angles, to trilateration, requiring the measurement of distances, was a marked shift in measurement-based strategies for estimating location, persistence of the prefix *tri* underscores the enduring importance of networks of triangles.
67. Deborah Jean Warner, "Political Geodesy: The Army, the Air Force, and the World Geodetic System of 1960," *Annals of Science* 59 (2002): 363–89. Warner describes an intense internal rivalry between Army geodesists and their counterparts in the Air Force.
68. Muneendra Kumar, "World Geodetic System 1984: A Modern and Accurate Global Reference Frame," *Marine Geodesy* 12 (1988): 117–26; and James R. Smith, *Introduction to Geodesy: The History and Concepts of Modern Geodesy* (New York: John Wiley & Sons, 1997), esp. 31–34.
69. Keith C. Clarke, "Geodesy," *The International Encyclopedia of Geography*, edited by Douglas Richardson and others (New York: John Wiley & Sons, 2017), DOI: 10.1002/9781118786352.wbieg0456. Clarke reported a revision of WGS84 in 2004. Also see D. Dong, T. Yunck, and M. Heflin, "Origin of the International Terrestrial Reference Frame," *Journal of Geophysical Research* 108, no. B4 (2003), 2200, DOI:10.1029/2002JB002035.
70. Clarke, "Geodesy"; and H. Moritz, "Geodetic Reference System 1980," *Journal of Geodesy*, 74 (2000): 128–162, DOI: 10.1007/S001900050278.
71. Larry Moore, "Datum Shifts and Digital Map Coordinate Displays," *US Geological Survey*, 2001, https://thor-f5.er.usgs.gov/drg/datum_shifts_v2.pdf; and US Geological Survey, "Topographic Map Margins Information for Educators," August 2013, http://education.usgs.gov/common/lessons/mapmargin.pdf.
72. Günter Seeber, *Satellite Geodesy*, 2nd ed. (Berlin: Walter de Gruyter, 2003), 147–57.
73. Seeber, 477.
74. Federal Aviation Administration, GPS Product Team, *Global Positioning System (GPS) Standard Positioning Service (SPS) Performance Analysis Report*, Global Positioning System (GPS) Standard Positioning Service (SPS) Performance Analysis Report no. 96, January 31, 2017, 10.

75. Thais C. Oliveira and Antonio F. B. A. Prado, "Mapping Orbits with Low Station Keeping Costs for Constellations of Satellites Based on the Integral over the Time of the Perturbing Forces," *Acta Astronautica* 104 (2014): 350–61.
76. US Air Force, 50th Space Wing Public Affairs, "50 SW to Dispose of Two GPS Satellites," March 16, 2016, http://www.schriever.af.mil/Public-Affairs/.
77. John Spencer, Brian G. Frizzelle, Philip H. Page, and John B. Vogler, *Global Positioning System: A Field Guide for the Social Sciences* (Malden, MA: Blackwell Publishing, 2003), esp. 27–33.
78. For a more exact description of the GPS timing system, see Seeber, *Satellite Geodesy*, 218–22.
79. Christophe Adrados, Irène Girard, Jean-Paul Gendner, and Georges Janeau, "Global Positioning System (GPS) Location Accuracy Improvement Due to Selective Availability Removal," *Comptes Rendus Biologies* 325 (2002) 165–70; and W. Y. Ochieng and K. Sauer, "Urban Road Transport Navigation: Performance of the Global Positioning System after Selective Availability," *Transportation Research Part C: Emerging Technologies* 10 (2002): 171–87.
80. Mark Monmonier, *Spying with Maps: Surveillance Technologies and the Future of Privacy* (Chicago: University of Chicago Press, 2002), 14–15.
81. Gordon West, "Differential GPS—How Accurate Is It?" *Trailer Boat* 25, no. 10 (1996): 72–74; and Adrados, Girard, Gendner, and Janeau, "Global Positioning System (GPS) Location Accuracy Improvement."
82. W. E. Larsen, G. A. Nielsen, and D. A. Tyler, "Precision Navigation with GPS," *Computers and Electronics in Agriculture* 11 (1994): 85–95. For discussion of the promise and questionable consequences of precision agriculture, see Monmonier, *Spying with Maps*, 47–49; and Alex McBratney, Brett Whelan, Tihomir Ancev, and Johan Bouma, "Future Directions of Precision Agriculture," *Precision Agriculture* 6, no. 1 (2005): 7–23.
83. For discussion of the reliability of driverless tractors, see Jacob Carballido del Rey, Juan Agüera Vega, Manuel Pérez-Ruiz, and Luis Emmi, "Comparison of Positional Accuracy between RTK and RTX GNSS Based on the Autonomous Agricultural Vehicles under Field Conditions," *Applied Engineering in Agriculture* 30 (2014): 361–66.
84. According to a discussion of accuracy posted on www.gps.gov, "high-quality" GPS receivers report their horizontal position to within 1.891 m (6.2 ft.) 95 percent of the time. Accuracy is affected by observing conditions. For example, if no buildings, bridges, or trees intervene, "GPS-enabled smartphones are typically accurate to within a 4.9 m (16 ft.) radius under open sky." US Air Force, National Coordination Office for Space-Based Positioning, Navigation, and Timing, "GPS Accuracy," accessed April 24, 2017, http://www.gps.gov/systems/gps/performance/accuracy/. For detailed accuracy data, see Federal Aviation Administration, GPS Product Team, "Global Positioning System (GPS) Standard Positioning Service (SPS) Performance Analysis Report." Elevation error might be twice as large.
85. US Coast Guard Navigation Center, Nationwide Differential GPS (NDGPS), accessed April 24, 2017, https://www.navcen.uscg.gov/.

86. "Wide Area Augmentation System (WAAS)," Targeted News Service, August 25, 2013, InfoTrac Newsstand, go.galegroup.com, accessed May 6, 2017. For an overview of current plans, see Federal Aviation Administration, " "Satellite Navigation-WAAS-How it Works," Accessed May 1, 2019, https://www.faa.gov/about/office_org/headquarters_offices/ato/service_units/techops/navservices/gnss/waas/howitworks/. Also see M. Bakri El-Arini, Walter Poor, Roland Lejeune, Robert Conker, James Fernow, and Kelly Markin, "An Introduction to Wide Area Augmentation System and Its Predicted Performance," *Radio Science* 36 (2001): 1233–40; Shau-Shiun Jan, "Vertical Guidance Performance Analysis of the L1-L5 Dual-Frequency GPS/WAAS User Avionics Sensor," *Sensors* 10 (2010): 2609–25; and "WAAS GPS Landing System Certified by FAA and Transport Canada," *Aircraft Engineering and Aerospace Technology* 81, no. 1 (2009): 261–62.
87. Joanne Will, "What's More Accurate: The Car's Speedo or the GPS?" *Globe and Mail* (Toronto), November 17, 2010.
88. Manuela Seitz et al., "The New DGFI-TUM Realization of the ITRS: DTRF2014 (data)," *Pangea*, 2016, DOI: 10.1594/PANGAEA.864046; and Technical University of Munich (TUM). "Positioning Exact to the Millimeter: Geodetic Reference System Enables Highly Accurate Positioning," *ScienceDaily*, August 25, 2016, www.sciencedaily.com/releases/2016/08/160825100943.htm.
89. Mohinder S. Grewal, Angus P. Andrews, and Chris G. Bartone, *Global Navigation Satellite Systems, Inertial Navigation, and Integration*, 3rd ed. (Hoboken, NJ: John Wiley & Sons, 2013), 4–10, 141–46. Also, at https://galileognss.eu/four-new-galileos-join-europes-largest-constellation/#more-3542, "The next Galileo launch is scheduled for 2020, the same year that Full Operational Capability is set to start."
90. Seeber, *Satellite Geodesy*, 15–17, 402–36, 485–99.
91. Josh Bavas, "Australia's Latitude and Longitude Coordinates Out by More than 1.5 Metres, Scientists Say," ABC News, July 27, 2016, http://www.abc.net.au/news/2016-07-28/aust-latitude-longitude-coordinates-out-by-1-5m-scientists/7666858; and Emily Clark, "Driverless Cars Need Australia's Latitude and Longitude Coordinates to Be Corrected," ABC News, July 28, 2016, http://www.abc.net.au/news/2016-07-28/why-it-matters-that-australias-coordinates-are-moving/7668014.
92. Chris Rizos and Donald Grant, "Australia on the Move: How GPS Keeps Up with a Continent in Constant Motion," *The Conversation*, February 5, 2017, https://theconversation.com/australia-on-the-move-how-gps-keeps-up-with-a-continent-in-constant-motion-71883.

Chapter 3

1. Edwin R. Stuart, *Topographical Drawing* (New York: McGraw-Hill, 1917), 68–71.
2. USGS manuals and web sources say little about the exact, standardized thickness of a topographic map's blue lines, or indeed about the intended width of any line on a printed USGS map. One precise width measurement I found was 0.15 mm (0.0059 in.), on appendix page II–19 in *Symbols for 1:50,000-Scale 15-minute Quadrangle Topographic Maps of the United States*, published jointly in October 1977 by the Geological Survey and the Defense Mapping

Agency (DMA). At that time, the Geological Survey and DMA were collaborating on standards for a map series covering foreign areas. Inspection of printed blue lines on a typical USGS topographic map using a ruler and magnifying lens indicated that 0.15 mm is a reliable estimate. A second source, from 2001, when printing was more precise, mentions a smaller line weight of 0.005 in. (0.127 mm) for streams and many other line features. See pp. 6–11 in *National Mapping Program Technical Instructions, Standards for 1:24,000- and 1:25,000-Scale Quadrangle Maps*, Part 6: Publication Symbols, US Geological Survey, July 2001.

3. Karen Severud Cook, "The Historical Role of Photomechanical Techniques in Map Production," *Cartography and Geographic Information Science* 29 (2002): 137–54. For discussion of water lining, see Stuart, *Topographical Drawing*, 68.
4. Although *atlas* commonly connotes a book of maps, the Geological Survey has used the term for a series of maps with a distinctive content, format, and scale. Individual map sheets were deemed akin to the pages in a bound atlas.
5. Institute for Government Research, *The US Geological Survey: Its History, Activities, and Organization* (New York: D. Appleton and Co., 1918), 14–15.
6. Donald C. Dahmann, "Gannett, Henry," *HC6*, 443–44.
7. Henry Gannett, *A Manual of Topographic Methods*, US Geological Survey Monograph 22 (Washington, DC, 1893).
8. Gannett, 41–76.
9. Gannett, 11.
10. Gannett, 11.
11. Gannett, 11.
12. Gannett, 12.
13. Gannett, 14.
14. Gannett, 107–8.
15. Gannett, 111–12. Gannett was writing about the more humid parts of the country, mostly east of the 30-in. rainfall line, and a more obvious focus of detailed topographic mapping.
16. For an overview of the Geological Survey's internal organization in the early twentieth century, see Institute for Government Research, *The US Geological Survey*, 73–84.
17. Gannett, *Manual of Topographic Methods*, 130.
18. Henry Gannett, *Manual of Topographic Methods*, US Geological Survey Bulletin 307 (Washington, DC, 1906).
19. *Topographic Instructions of the United States Geological Survey* (Washington, DC, 1913). No contributors are named, but a prefatory note by Robert Bradford Marshall indicates that the various sections were "prepared by the members of the topographic branch best qualified for the task" (9). Symbol plates are on 206–25, lettering examples are on 226 and 227, and the abbreviations are on 228.
20. *Topographic Instructions*, 205.
21. The executive order was signed on August 10, 1906, and rescinded on December 30, 1919. *Report of the United States Geographic Board* [5th Report of the United States Board on Geographic Names] (Washington, DC, 1920), 6.

22. War Department Document No. 418 is mentioned by Stuart, *Topographical Drawing*, v and 51; by J. K. Finch, in his *Topographic Maps and Sketch Mapping* (New York: John Wiley & Sons, 1920), 8; and by War Department Bulletin No. 64 (Washington, DC, December 19, 1918). The symbols were published as a booklet and a 20 × 30 in. map-size sheet, both printed by the Geological Survey. See War Department, Office of the Chief of Staff, *Conventional Signs: United States Army Maps*, War Department Document No. 418 (Washington, DC, 1912), and US Geological Survey, "Conventional Signs Adopted by the United States Geographic Board" (chart) (Washington, DC, 1911).
23. *Topographic Instructions of the Geological Survey* (1913), 205.
24. *Topographic Instructions*, 163.
25. *Topographic Instructions*, 168.
26. For a concise overview of geometry of surface networks, see Sanjay Rana, *Topological Data Structures for Surfaces: An Introduction to Geographical Information Science* (Chichester, UK: John Wiley & Sons, 2004), esp. 52–58.
27. *Topographic Instructions of the Geological Survey* (1913), 148.
28. *Topographic Instructions*, 159.
29. *Topographic Instructions*, 162–63.
30. *Topographic Instructions*, 159.
31. *Topographic Instructions*, 208.
32. *Topographic Instructions*, 138.
33. New York State voters approved the Barge Canal in a 1903 referendum. Work was completed in 1918. General references are Noble E. Whitford, *History of the Barge Canal of New York State* (Albany, NY: J. B. Lyon, 1922); and Alan Scot Willis, "Erie Canal," in *Encyclopedia of New York*, ed. Peter Eisenstadt (Syracuse, NY: Syracuse University Press, 2005), 514–22.
34. Noble E. Whitford, *History of the Canal System of the State of New York* (Albany, NY: Brandow Printing Co., 1906), 1042.
35. Downtown Syracuse had another kind of lock, not shown on the map. This was a weigh lock, in a structure now known as the Weighlock Building and home of the Erie Canal Museum. Like other locks, a weigh lock had gates at both ends. It also had a scale at the bottom onto which a boat was lowered when the water was removed. Weighing an unladen boat provided a registered tare weight that could be subtracted from the gross weight of the loaded boat to calculate the weight of its cargo. Martin Morganstein and Joan H. Cregg, *Erie Canal* (Charleston, SC: Arcadia Pub. Co., 2001), 21–22, 56–57, 61.
36. Although the canal's contour crossing on the right of the final *E* in "ERIE" might suggest that water in the canal was flowing west at this point, the topographer was probably trying to portray subtle variations in slope in the largely level and nearly flat landscape onto which the canal had been imposed. Canals were inherently level structures, excavated in places but supported by embankments in others.
37. Whitford, *History of the Canal System*, 1104.
38. See, for example, Mary Beth Norton and others, *A People and a Nation: A History of the United States* (Mason, OH: Cengage Learning, 2014), 281.

39. DeRuyter Reservoir was completed in 1863. Whitford, *History of the Canal System*, 1104.
40. The Old Erie Canal website, maintained by Frank E. Sadowski Jr., has photographs of the aqueduct as well as overhead images from Google Maps; see "Limestone Creek Aqueduct," accessed May 26, 2017, http://www.eriecanal.org/LimestoneCreek.html.
41. *Carhart v. State of New York*, 115 App. Div. 1 (N.Y. App. Div. 1906).
42. Thomas K. Grasso, "Oswego Canal," *Encyclopedia of New York*, ed. Peter Eisenstadt (Syracuse, NY: Syracuse University Press, 2005), 1164; Field Horne, "Canals," *Encyclopedia of New York*, ed. Peter Eisenstadt (Syracuse, NY: Syracuse University Press, 2005), 256–57; Wilfred H. Schoff, "The New York State Barge Canal [part 2]," *Bulletin of the American Geographical Society* 47 (1915): 498–508; and Whitford, *History of the Canal System*, 1466.
43. Whitford, *History of the Barge Canal*, 169, 471, 488.
44. The vertical line on the right in figure 3.7 is the meridian 76°30′ W, which separates the Oswego, N.Y., and Fulton, N.Y., 15-minute quadrangles and cuts through the eastern part of the city. Around 1904, the Geological Survey published the *Oswego Special Map, NY*, a 15-minute quadrangle map, which keeps the city intact by incorporating parts of the Oswego and Fulton sheets. The Geological Survey published a similar special sheet that kept Rochester, New York, intact as well as the aptly named *Rochester Long* sheet, on which a greater than usual vertical extent accommodated the entire city.
45. New York Department of State Engineer and Surveyor, *Report on Hydraulic Data*, 1916, Supplement to the Annual Report of the State Engineer and Surveyor for the Fiscal Year Ended June 30, 1916 (Albany, NY: J. B. Lyon Co., 1917), 79.
46. Railroad mileages are from the *Official Guide of the Railways and Steam Navigation Lines of the United States* 90 (June 1957): 169, 182.
47. During World War II, Fort Ontario was a haven for Jewish refugees from Nazi persecution. For critical history of the shelter, see Sharon R. Lowenstein, *Token Refuge: The Story of the Jewish Refugee Shelter at Oswego, 1944–1946* (Bloomington: Indiana University Press, 1986).
48. Felix E. Reifschneider, *Trolley Lines of the Empire State (City and Suburban)* (Orlando, FL, 1960), 36; and William R. Gordon and James R. McFarlane, *The Rochester, Syracuse, and Eastern "Travelectric" 1906–1931* (Rochester, NY, 1961), 155–56.
49. A map's reprint date is disclosed in the lower-right corner, just below the neatline. Reprint dates for the Oswego sheet are from the collection of digital editions scanned by the Geological Survey and acquired by John Olson, map librarian at Syracuse University. The Geological Survey did not revise its 15-minute Oswego, New York, map until 1960, when it issued a substantially updated version, compiled from a quartet of more detailed 7.5-minute maps, based partly on aerial photography taken in 1942.
50. *Topographic Instructions of the United States Geological Survey* (Washington, DC, 1913), 207.
51. US Geological Survey, *Lettering and Conventional Signs*, edition of November 1898.
52. Mark Monmonier, *Mapping It Out: Expository Cartography for the Humanities and Social Sciences* (Chicago: University of Chicago Press, 1993), 67–68.

53. See, for example, plate *O*, parts of Wards 3, 5, and 7, in G. M. Hopkins, *City Atlas of Oswego, New York* (Philadelphia, 1880), online at the Library of Congress, accessed June 15, 2017, http://hdl.loc.gov/loc.gmd/g3804om.gla00259.
54. US Geological Survey, 1901, Chicago, Illinois, 1901, 15-minute quadrangle map.
55. "O&W Railroad Pedestrian Promenade and Bikeway Facts," Rails-to-Trails Conservancy, accessed June 15, 2017, https://www. traillink.com/trail/ow-railroad-pedestrian -promenade-and-bikeway/.
56. *Topographic Instructions of the United States Geological Survey* (Washington, DC, 1913), 139.
57. A University of Chicago geology professor and a map librarian who collaborated on a cross-national study of topographic map symbols found only "a moderate standardization on the symbols used for railroads" but observed that "minor differences which may be encountered will seldom prove confusing." See Everett C. Olson and Agnes Whitmore, *Foreign Maps* (New York: Harper and Bros., 1944), 147.
58. *Topographic Instructions of the United States Geological Survey* (Washington, DC, 1913), 138; and *Topographic Instructions* (Washington, DC, 1918), 149.
59. W. M. Beaman, *Topographic Mapping*, part E of C. H. Birdseye, *Topographic Instructions of the United States Geological Survey*, Bulletin 788 (Washington, DC, 1928), 286.
60. Although the author diagram in figure 3.2 indicates that E. B. Clark had primary responsibility for the part of the quadrangle map around the city of Oswego—the part that includes areas shown in figures 3.6 and 3.7—my use of the plural topographers assumes a modest consultation between Clark and coauthor James McCormick. Moreover, according to the *Topographic Instructions* for 1913, "before being transmitted to the inspectors for examination, the original drawing, together with all auxiliary map data .... must be submitted, through the division geographer, to a disinterested topographer for rigorous proof reading as to its completeness, accuracy, and general conformity with Survey instructions" (171).
61. Network features and structures were traced in Adobe® Illustrator® to avoid graphic complications related to oxidizing paper, color symbols, and parts of the images with dark relief shading. For clarity, the drawing also omits elevation contours and isobaths but reproduces the style and placement of labels.
62. According to the page of Kodtrak Kountry's rail fan website, the roundhouse was part of New York Central's West Yard, noteworthy for its 692-car capacity facility with a roundhouse that once had thirty stalls (but was reduced to ten by the 1940s), a wye and complete servicing facilities," http://kodtrak.railfan.net/oswego%20nyc.html (accessed June 17, 2017).
63. Joseph A. Kotch Jr., "The 1:250,000-Scale Map Series of the United States" (MA paper, Syracuse University, 1982), 22; and Morris Mordecai Thompson, *Maps for America: Cartographic Products of the US Geological Survey* (Reston, VA, 1979), 32 and 114.
64. The railroad symbol on the 1965 map is noticeably thicker than the sample in Army Map Service, US Navy Hydrographic Office, and Aeronautical Chart and Information Center, *Joint Specifications for Map and Chart Compilation, Scale 1:250,000* (1955), 147. Comparison with the immediate post–World War II edition (1948 in the case of the Rochester, NY, sheet) indicates that the width of the line symbol for a single-track railroad increased by at least three times.

65. Thompson, *Maps for America*, 52.
66. US National Resources Committee, *Suggested Symbols for Plans, Maps, and Charts* (Washington, DC, June 1938). A preliminary edition published in 1936 was followed by confirmed editions in 1937 and 1938.
67. "Court Approves Plan for Settling Claims Against Defunct O. & W.," *New York Times*, January 20, 1960, 41; and William F. Helmer, *O. & W.: The Long Life and Slow Death of the New York, Ontario & Western Railway* (Berkeley, CA: Howell-North, 1959), esp. 160–67.
68. "Central Offers Help: Will Provide Some Switching Service to Replace O. & W.," *New York Times*, April 3, 1957, 53.
69. James William Darlington, "A Railroad Geography: The New York, Ontario, and Western Railway" (MA thesis, Syracuse University, 1974), esp. 133–36. The O&W map in figure 3.13 is from "New York, Ontario, and Western Railway," *Official Guide of the Railways and Steam Navigation Lines of the United States*, June 1957, 168–69. In addition to declaring "This road discontinued operation March 29, 1957," the listing mentioned General Offices, at 39 Broadway, NYC, used bold type to identify "Jacob Grumet and James B. Kilsheimer, III, Receivers, New York, NY," and left vacant the name slot for "Chairman of the Board and President."
70. Jerry Musich, "Mapping a Transcontinental Nation: Nineteenth- and Early Twentieth-Century American Rail Travel Cartography," in *Cartographies of Travel and Navigation*, ed. James R. Akerman (Chicago: University of Chicago Press, 2006), 97–150; and Mark Ovenden, "Railroad Map," *HC6*, 1239–41.
71. Karen Severud Cook, "Wax Engraving," *HC6*, 1701–4; and David Woodward, *The All-American Map: Wax Engraving and Its Influence on Cartography* (Chicago: University of Chicago Press, 1977), 50–93.
72. Woodward, *All-American Map*, 38–39.
73. Woodward, 85.
74. John H. Long, "London Underground Map," *HC6*, 788–92.
75. Gerald A. Danzer, "Wayfinding and Travel Maps: Public Transportation Map," *HC6*, 1727–33.
76. Daniel Dorling, "Cartogram," *HC6*, 193–97.
77. Ralph E. Ehrenberg, "Airline Map," *HC6*, 35–40, esp. 39.

Chapter 4

1. Although claims range from thirty to fifty tons for a horse working a canal, writers seem to agree on two tons per horse along what passed for a "surfaced" road in post-colonial America. For examples, see Christopher F. Jones, *Routes of Power: Energy and Modern America* (Cambridge, MA: Harvard University Press, 2014), 28; and William A. Newman and Wilfred E. Holton, *Boston's Back Bay: The Story of America's Greatest Nineteenth-Century Landfill Project* (Hanover, NH: University Press of New England, 2006), 79–80.

2. George Geddes, "The Erie Canal: Origin and History of the Measures That Led to Its Construction," *Publications of the Buffalo Historical Society* 2 (1880): 263–304, esp. 269.
3. Gerard Koeppel, *Bond of Union: Building the Erie Canal and the American Empire* (Boston: Da Capo Press, 2009), 21.
4. The nation's three early canals were the Dismal Swamp Canal, in Virginia; the Santee Canal, in South Carolina; and the Middlesex Canal, in Massachusetts. Ronald E. Shaw, *Canals for a Nation: The Canal Era in the United States, 1790–1860* (Lexington, KY: University of Kentucky Press, 1990), 9–22, 117–20.
5. Pamela L. Baker, "The Washington National Road Bill and the Struggle to Adopt a Federal System of Internal Improvement," *Journal of the Early Republic* 22, no. 3 (2002): 437–64.
6. Harmeet Sawhney, "Dynamics of Infrastructure Development: The Role of Metaphors, Political Will, and Sunk Investment," *Media, Culture & Society* 23 (2001): 33–51, esp. 38–39.
7. James Geddes, "Report of James Geddes to the Surveyor General, with Appendices A, B, C, in 1809, on the Inland Navigation between Hudson's River and Lake Erie, Made Pursuant to Concurrent Resolutions of the Senate and Assembly of 21st March and 6th April, 1808," in New York [State], Secretary of State, *Laws of the State of New York, in Relation to the Erie and Champlain Canals: Together with the Annual Reports of the Canal Commissioners, and Other Documents Requisite for a Complete Official History of Those Works* (Albany, NY: E. and E. Hosford, 1825), 13–32, quotations on 31.
8. Geddes, "Report," 13.
9. *Gerundegut* is a variant of Irondequoit, a native American name more common on present-day maps, which show Irondequoit Bay, Irondequoit Creek, and the Town of Irondequoit. Respective translations include "where the land and waters meet" and "the place where the waves gasp and die." See Thomas Cole Richens and Patricia S. Wayne, *Irondequoit* (Charleston, SC: Arcadia Publishing, 2006), 7; and Ren Vasiliev, *From Abbotts to Zurich: New York State Placenames* (Syracuse: Syracuse University Press, 2004), 114.
10. James Geddes, "Letter from James Geddes Esqr., Canal Engineer, to Mr. William Darby, Albany, February 22, 1822," in *Laws of the State of New York, in Relation to the Erie and Champlain Canals*, 42–46, quotations on 43–44.
11. Geddes to Darby, 43.
12. Geddes to Darby, 44.
13. Noble E. Whitford, *History of the Canal System of the State of New York* (Albany, NY: Brandow Printing Co., 1906), 794. Books that endorse the 70-ft. elevation include F. Daniel Larkin, *New York State Canals: A Short History* (Fleischmanns, NY: Purple Mountain Press, 1998), 20; Ronald E. Shaw, *Erie Water West: A History of the Erie Canal, 1792–1854* (Lexington, KY: University of Kentucky Press, 1966), 404; and Shaw, *Canals for a Nation*, 39.
14. Whitford, *History of the Canal System*, 104.
15. Tim McNeese, *The Erie Canal: Linking the Great Lakes* (New York: Chelsea House, 2009), 91–92. A similar narrative can be found in Ralph K. Andrist, *The Erie Canal* (Columbus, GA: American Heritage/New Word City, 2016), 48–49. According to Shaw, who qualified

"the tradition that Irish workers were recruited abroad to dig the canal," Irish and Welsh laborers were not numerous until after 1821, when work on the Gerundegut Embankment was starting. Shaw, *Canals for a Nation*, 38.

16. Jenny Marsh Parker, *Rochester: A Story Historical* (Rochester, NY: Scrantom, Wetmore and Co., 1884), 115.
17. In showing only a single canal along the Mohawk, I was leery of whatever confusion affected the canal map in the *Encyclopedia of New York State*, in which the symbols (or labels) for "Barge Canal" and "Clinton's Ditch and first enlargement" are oddly reversed. See "Erie Canal," in *Encyclopedia of New York State* (Syracuse: Syracuse University Press, 2010), 514–20, map on 515. Rare blunders such as this one are a warning to cartographers, map editors, and authors to check all symbols and labels.
18. Whitford, *History of the Canal System*, 1030, 1035. For a concise critique of the Black River Canal, described as "successful only in its engineering," see Shaw, *Canals for a Nation*, 45–46, quotation on 46. Also see Edward P. Fynmore and Harney J. Corwin, *Black River Canal*, Images of America series (Charleston, SC: Arcadia Publishing, 2005).
19. Wilfred H. Schoff, "The New York State Barge Canal (part II)," *Geographical Review* 47 (1915): 498–508, esp. 504–5.
20. Gerard Koeppel, *Bond of Union: Building the Erie Canal and the American Empire* (Boston: Da Capo Press, 2009), 9.
21. "Break in the New York State Barge Canal at the Irondequoit Creek Crossing," *Engineering News* 68, no. 12 (September 19, 1912): 548–54. Also see "Repairing the Break in the Irondequoit Embankment on the Erie Canal," *Engineering News* 68, no. 14 (October 3, 1912): 548–54.
22. According to the US Inflation Calculator (www.usinflationcalculator.com), which adjusts for inflation over the period 1913 to 2017, $200,000 in 1913 would purchase goods priced at $4,983,090.91 in 2017. Accessed December 12, 2017.
23. Doug Farley, "Canal Discovery: The Irondequoit Embankment," *Lockport (NY) Union-Sun & Journal Online*, February 20, 2010.
24. The Canandaigua Outlet is a river that was altered to provide a link between the Erie Canal and Canandaigua Lake. This river joined Ganargua Creek to form the Clyde River.
25. National Park Service, "Erie Canalway" (Washington, DC: US Department of the Interior, 2006); and Wayne County Planning Department, *Wayne County Parks and Recreationways Master Plan* (Lyons, NY, 2012), maps in appendix A between pp. 34 and 35.
26. Parks & Trails New York, *Cycle the Erie Canal* (Interactive Trail Map), https://www.ptny.org/bike-canal/map/.
27. "Canal System No Liability," *Albany (NY) Times Union*, September 30, 2016, C1.
28. "Tolls, Passes and Permits," New York State Canal Corp., accessed December 14, 2017, http://www.canals.ny.gov/boating/tolls.html. Annual fees for commercial vessels ranged from $750 for tug boats and barges to "$10 per passenger capacity, for hour dayliners." "Commercial Shipping and Towing," New York State Canal Corp., accessed December 14, 2017, http://www.canals.ny.gov/business/shipping.html.

29. In 2015, seventeen gaps totaling seventy-two miles had yet to be eliminated, but funding had already been arranged for seven of them. "Closing the Gaps: A Progress Report on the Erie Canalway Trail 2015," Parks & Trails New York, accessed December 14, 2017, https://www.ptny.org/application/files/8714/5935/3322/2015-Closing-the-Gaps-Report.pdf.
30. Richard F. Palmer, "Abandoned Railroads in New York State," *National Railway Historical Society Bulletin* 45, no. 5 (1980): 8–21, 36–39, esp. 37.
31. John F. Stover, *American Railroads*, 2nd ed. (Chicago: University of Chicago Press, 1997), 205. Comparable summary data for the US canal network are scarce. For 1880, Michael Mulhall estimated a total of 4,479 mi. of canal, of which 1,964 were reported as abandoned. For that year, he also reported 47,355 mi. of navigable rivers, and for 1860, he attributed the 5,460 mi. of canals in use to 118 canals. See Michael G. Mulhall, *The Dictionary of Statistics* (London: George Routledge and Sons, 1892), 105. Mulhall's data were accepted as authoritative and suitably current in A. Barton Hepburn, *Artificial Waterways and Commercial Development (with a History of the Erie Canal)* (New York: The Macmillan Co., 1909), 13–14.
32. Raymond's experience in railroading began in 1879, at the Kansas Pacific Railway, which the Union Pacific bought in 1880, and included summer employment at the Kansas City, Fort Scott & Gulf Railroad and a year (1882–83) as an assistant division engineer for the Kansas City, Springfield & Memphis Railroad. After graduation from Washington University, he worked for six years as an engineering instructor at the University of California while consulting with railways in the Mississippi Valley on location and construction. In 1892, after two years in private practice, he was appointed Professor of Geodesy, Road Engineering, and Engineering Drawing at Rensselaer Polytechnic Institute, in Troy, New York, where he consulted for local industries on water power projects. He also served as chief engineer for both the city water system and the Troy & New England Electric Railway, a local trolley line. *National Cyclopaedia of American Biography* 17 (1921): 427–28; and Prabook [online encyclopedia], s.v. "William Galt Raymond," accessed December 18, 2017, https://prabook.com/web/william_galt.raymond/1050935.
33. William G. Raymond, *Elements of Railroad Engineering* (New York: John Wiley & Sons, 1908). Subsequent editions were issued in 1913, 1917, 1923, 1937, and 1949; the last two, published posthumously, listed multiple coauthors.
34. This paragraph condenses and generalizes the more exact description in Raymond, *Elements*, 19–22.
35. For an extensive discussion of rise and fall, see Raymond, *Elements*, 217–29.
36. Raymond, *Elements*, 250.
37. Raymond, *Elements*, 244.
38. Raymond, *Elements*, 243.
39. Raymond, *Elements*, 244.
40. Thomas M. Cleemann, *The Railroad Engineer's Practice, Being a Short but Complete Description of the Duties of the Young Engineer in Preliminary and Location Surveys and in Construction* (New York: D. Van Nostrand, 1892), map between 20 and 21; quotation on 21. Cleemann attributed the map to H. Eaton Coleman, presumably a railroad employee or

consultant. The map was added to the 1892 edition of the textbook, first published in 1880 in New York City by George H. Frost. In November 1893, Cleemann died in Ecuador, at age 50, after a week-long fever. His brief obituary noted that he held engineering degrees from the University of Pennsylvania and Rensselaer Polytechnic Institute and had been in Ecuador to advise on "some important public works." "Tete-a-Tete" [column], *Philadelphia Times*, November 18, 1893, 4.

41. Raymond, *Elements*, 181–84, 336–40.
42. Raymond, *Elements*, 19–20, quotation on 20.
43. William Beahan, *Field Practice of Railway Location* (New York: Engineering News, 1904).
44. For examples, see Howard Ande, "Tehachapi in the 21st Century," *National Railway Historical Society Bulletin* 75 (Spring 2010): 4–21.
45. Beahan, *Field Practice*, 57.
46. Brett Hansen, "Completing the Loop: The Tehachapi Pass Railroad," *Civil Engineering* 80, no. 4 (April 2010): 38–39. Also see, Historical Marker Database, "Tehachapi Pass Railroad Line, Constructed 1874–1876, Commemorated October 1998," https://www.hmdb.org/marker.asp?marker=11910; American Society of Civil Engineers, "Historic Landmarks," www.asce.org/landmarks/; American Society of Civil Engineers, "Tehachapi Pass Railroad Line," http://www.asce.org/project/tehachapi-pass-railroad-line/.
47. John F. Stover, *American Railroads*, 2nd ed. (Chicago: University of Chicago Press, 1961, 1997), esp. 164–65; and George W. Hilton and John F. Due, *The Electric Interurban Railways in America* (Stanford, CA: Stanford University Press, 1960), esp. 5–15.
48. James R. Akerman, "Road Mapping in Canada and the United States," in *HC6*, 1339–50; and Peter J. Hugill, "Good Roads and the Automobile in the United States, 1880–1929," *Geographical Review* 72 (July 1982): 327–49.
49. The Federal Highway Administration, now in the US Department of Transportation, maintains the timeline "Milestones for US Highway Transportation and the Federal Highway Administration," at https://www.fhwa.dot.gov/publications/publicroads/96spring/p96sp44.cfm. Also useful is the agency's General Highway History web page, at https://www.fhwa.dot.gov/highwayhistory/history_misc.cfm.
50. Richard F. Weingroff, "Federal Aid Road Act of 1916: Building the Foundation," *Public Roads* 60, no. 1 (1996), online at https://www.fhwa.dot.gov/publications/publicroads/96summer/p96su2.cfm (accessed December 26, 2017); and Weingroff, "Federal-Aid Highway Act of 1956: Creating the Interstate System," *Public Roads* 60, no. 1 (1996), online at https://www.fhwa.dot.gov/publications/publicroads/96summer/p96su10.cfm.
51. Walter Wilson Crosby, "Preliminary Investigations," in *American Highway Engineers' Handbook*, ed. Arthur H. Blanchard (New York: John Wiley & Sons, 1919), 147–92, quotations on 148.
52. T. F. Hickerson, "Road Surveying and Mapping," *Proceedings of the Good Roads Institute*, University of North Carolina, March 17–19, 1914, North Carolina Geological and Economic Survey, economic paper no. 39, 19–26, quotation on 19.
53. In the early 1950s, the Highway Research Board systematically surveyed state highway departments to ask about strategies for increasing efficiency. As the summary reported,

"Among the fast developing techniques is that of aerial photography [in] use by 47 Planning Bureaus, 46 Locating Bureaus, 29 Design Bureaus, 21 Materials and Tests Bureaus, 18 Soils Bureaus, and a substantial number of Right-of-Way Bureaus." The report concluded, "In many highway departments, the aerial survey method has become the standard method for preliminary location and mapping with ground surveys limited to control and spot checking of detail." US National Research Council, Highway Research Board, "Time-Saving Methods in Highway Engineering–Chapter 14: Summary, Selected References, and Subject Index" (Washington, DC, 1956), quotations on 363; online at http://pubsindex.trb.org/view/1956/m/132294.

54. Patrick H. McHaffie, "US Geological Survey," in *HC6*, 1659–66; Gottfried Konecny, "Analytical Photogrammetry and Control Surveying," in *HC6*, 1125–32; and James Reilly, "Geodesy and Photogrammetric Mapping," in *HC6*, 1132–36.
55. William T. Pryor, "The Role of Aerial Surveys in Highway Engineering" (paper presented at the Ninth Congress, International Society for Photogrammetry, London, England, September 5–17, 1960), Hathi Trust (catalog.hathitrust.org/Record/100648229).
56. Mark Monmonier, *Rhumb Lines and Map Wars: A Social History of the Mercator Projection* (Chicago: University of Chicago Press, 2004), 98–109.
57. Hugh C. Mitchell and Lansing G. Simmons, *The State Coordinate Systems (A Manual for Surveyors)*, US Coast and Geodetic Survey Publication no. 235 (Washington, DC, 1945), vi.
58. Joel L. Morrison, "Coordinate Systems," in *HC6*, 278–84, esp. 279–81.
59. See, for example, Louis Bell, *Electric Power Transmission: A Practical Treatise for Practical Men*, 5th ed. (New York: McGraw Publishing Co., 1907), 389; and Rollin W. Hutchinson Jr., *Long-Distance Electric Power Transmission, Being a Treatise on the Hydro-Electric Generation of Energy; Its Transformation, Transmission, and Distribution* (New York: D. Van Nostrand Co., 1907). Although Bell found USGS maps "invaluable, when available, for getting a preliminary idea of the topography and the probable drainage basin," neither book, despite promising titles, mentioned the use of maps for locating transmission lines. No less dismissive, except for assessing drainage area (as an indicator of hydro-electric generating potential), are Frank Koester, *Hydroelectric Developments and Engineering: A Practical and Theoretical Treatise on the Development, Design, Construction, Equipment, and Operation of Hydroelectric Transmission Plants*, 2nd ed. (New York: D. Van Nostrand, 1911), 12–14; and H. A. E. C. von Schon, *Hydro-Electric Practice: A Practical Manual of the Development of Water Power, Its Conversion to Electric Energy, and Its Distant Transmission* (Philadelphia: J. B. Lippincott Co., 1908), esp. 26–28.
60. Alfred Still, *Overhead Electric Power Transmission: Principles and Calculations* (New York: McGraw-Hill, 1913), quotations on 4–5.
61. Ruben A. Lundquist, *Transmission Line Construction: Methods and Cost* (New York: McGraw-Hill, 1912), quotations on 1–2.
62. Matthews-Northrup Works, "Map Showing the Electric Power Lines in the State of New York Issued by the Niagara Lockport and Ontario Power Company," Buffalo, NY, 1900–1950, catalog entry accessed December 29, 2017, http://summit.syr.edu/vwebv/holdingsInfo?bibId=4640177.

63. Date from photographs of the construction of the Salmon River Dam; see *Half-Shire Historical Society, Northern Oswego County* (Charleston, SC: Arcadia Publishing Co., 2003), 54–55.
64. US Geological Survey, 1905, Orwell, New York, 15-minute quadrangle map; US Geological Survey, 1895, Pulaski, New York, 15-minute quadrangle map; US Geological Survey, 1906, Kasoag, New York, 15-minute quadrangle map; and US Geological Survey, 1905, Mexico, New York, 15-minute quadrangle map.
65. US Geological Survey, 1898, Syracuse, New York, 15-minute quadrangle map.
66. Carl F. Meyer, *Route Surveying* (Scranton, PA: National Textbook Co., 1949), 269–71, quotation on 270.
67. Meyer, *Route Surveying*, 8.
68. Meyer, *Route Surveying*, 9.
69. Meyer, *Route Surveying*, chap. 11.
70. The scheme also includes purple for "reclaimed water, irrigation, and slurry lines" and pink for "temporary survey markings." See American Public Works Association, "APWA Uniform Color Code" (Washington, DC, 1999), accessed December 31, 2017, http://www2.apwa.net/documents/About/TechSvcs/One-Call/COLORCC.PDF.

Chapter 5

1. Erik Larson, *Isaac's Storm: A Man, a Time, and the Deadliest Hurricane in History* (New York: Random House, 1999). The book's title commemorates local meteorologist Isaac Cline, forced by Weather Bureau officials in Washington to ignore forecasts by Cuban meteorologists, who had predicted a massive storm headed his way—an early twentieth-century example of scientific nationalism. For discussion of American efforts to establish a telegraphic observer network in the West Indies in the aftermath of the Spanish-American War and disagreements with native meteorologists, see Jamie L. Pietruska, "Hurricanes, Crops, and Capital: The Meteorological Infrastructure of American Empire in the West Indies," *Journal of the Gilded Age and Progressive Era* 15 (2016): 418–45, esp. 430–32.
2. Breslau was in the province of Silesia, in the German state of Prussia. Most of Silesia became part of Poland after World War II. Heavily damaged by bombing, the University of Breslau was rebuilt and renamed the University of Wrocław. Little remains, architecturally or in Polish history, of the university's roots. Gregor Thum, *Uprooted: How Breslau Became Wrocław during the Century of Expulsions* (Princeton, NJ: Princeton University Press, 2003), xii–xxxi, 395.
3. Heinrich Wilhelm Brandes, "Aus einem Schreiben des Professor Brandes, meteorologischen Inhalts" (from Professor Brandes's letter of meteorological content), *Annalen der Physik* 55 (1817): 112–14. For a translation, see Karl Schneider-Carius, *Wetterkunde, Wetterforschung: Geschichte ihrer Probleme und Erkenntnisse in Dokumenten aus drei Jahrtausenden (Weather Science, Weather Research: History of Their Problems and Findings from Documents During Three Thousand Years)* (New Delhi: Indian National

Scientific Documentation Centre; Washington, DC: National Oceanic and Atmospheric Administration and the National Science Foundation, 1975), 179–80. For discussion of Brandes's project and its impact, see Mark Monmonier, *Air Apparent: How Meteorologists Learned to Map, Predict, and Dramatize Weather* (Chicago: University of Chicago Press, 1999), 18–22, 239n4, 239n5.

4. Brandes, "Einige Resultate aus der Witterungs-Geschichte des Jahres 1783, und Bitte um Nachrichten aus jener Zeit" ("Some Results from the Weather History of the Year 1783, and a Request for News from that Time"), *Annalen der Physik* 61 (1819): 421–26.
5. Wilhelm Trabert, *Meteorologie und Klimatologie* (Leipzig: F. Deuticke, 1905), 65. For a similar but not identical pattern of isobars for the same day, see H. Hildebrand Hildebrandsson and Léon Philippe Teisserenc de Bort, *Les bases de la météorologie dynamique historique—* ("Elements of the history of dynamic meteorology—state of knowledge") (Paris: Gauthier-Villars et fils, 1907), 1:47.
6. See the entry "Isobar," in Helen Wallis and Arthur H. Robinson, *Cartographical Innovations: An International Handbook of Mapping Terms to 1900* (Tring, Herts, UK: Map Collector Publications in association with the International Cartographic Association, 1987), 223–24. Wallis and Robinson credit Brandes with the first cartographic application of measurements based on the barometer, invented by Evangelista Torricelli "in the mid-17th century [though] variations from place to place were not mapped until 1820," a date that seems off by at least a year.
7. David C. Cassidy, "Meteorology in Mannheim: The Palatine Meteorological Society, 1780-1795," *Sudhoffs Archiv* 69 (1985): 8–25.
8. Cassidy, "Meteorology in Mannheim," 23–25.
9. John Kington, *The Weather of the 1780s over Europe* (Cambridge: Cambridge University Press, 1988), 5–11.
10. Cassidy, "Meteorology in Mannheim," 19–20.
11. Cassidy, "Meteorology in Mannheim," 21–22.
12. Kington, *The Weather of the 1780s*, 6–11; Cassidy, "Meteorology in Mannheim"; and John Frederic Daniell, "On the Oscillations of the Barometer," *Quarterly Journal of Science and the Arts* 21 (1826): 82–100.
13. Monmonier, *Air Apparent*, 39–54.
14. For an overview of Joseph Henry's career, see Thomas Coulson, *Joseph Henry: His Life and Work* (Princeton, NJ: Princeton University Press, 1950); Albert E. Moyer, *Joseph Henry: The Rise of an American Scientist* (Washington, DC: Smithsonian Institution Press, 1997); and the "Joseph Henry Timeline" posted online by the Smithsonian Institution, at https://siarchives.si.edu/timeline/featured-topics/henry.
15. Franklin R. Hough, *Results of a Series of Meteorological Observations Made in Obedience to Instructions from the Regents of the University at Sundry Academies in the State of New York, from 1826 to 1850, Inclusive* (Albany: Weed, Parsons, and Co., 1855).
16. Simeon DeWitt also held the title Chancellor of the University of the State of New York, not to be confused with today's State University of New York (SUNY), established in 1948.

17. For Henry's recollection of his telegraphic bell-ringing experiment at the Albany Academy and a fellow scientist's corroboration, see *Annual Report of the Board of Regents of the Smithsonian Institution, Showing the Operations, Expenditures, and Condition of the Institution for the Year 1859* [hereafter *Smithsonian Report*], 96, 104–6.
18. Coulson, *Joseph Henry*, 108, 120; Moyer, *Joseph Henry*, 143–44; and Robert Luther Thompson, *Wiring a Continent: The History of the Telegraph Industry in the United States, 1832–1866* (Princeton, NJ: Princeton University Press,1947), 4–5.
19. As an example, for 1849 the "Tabular View of Public Libraries in the United States" lists numbers of libraries and volumes held, by state and apportioned among seven categories of library. *Smithsonian Report, 1849*, 189–91.
20. Commercial ventures had been collecting weather reports by telegraph from selected cities as early as 1848. James Rodger Fleming, *Meteorology in America, 1800–1870* (Baltimore: Johns Hopkins University Press, 1990), 142–43.
21. *Smithsonian Report*, 1856, 35.
22. *Smithsonian Report*, 1858, 33.
23. Smithsonian Institution, *Map Showing the Stations Where Meteorological Observations Are Made in the United States under the Direction of the Smithsonian Institution [in] 1851,* scale 1:6,500,000 (Washington, DC, 1851). The Library of Congress has two copies, mounted on linen and possibly acquired in the early twentieth century. Both lack a stamped date and accession number. The catalog record is online at https://lccn.loc.gov/98686197. Smithsonian authorship is inferred from the title rather than explicitly stated. The excerpt in figure 5.4 was rotated clockwise 5.4°, enlarged slightly, and adjusted in Photoshop to improve the clarity of labels and symbols.
24. The length of a 60-mi. bar scale was estimated along the parallel 42° N (roughly the horizontal southern border with Pennsylvania) between the southwestern corner of New York State and the state's eastern border with Connecticut.
25. *Smithsonian Report*, 1851, 70–71, and 1852, 75–76. My hunch is that italic type was used, with questionable consistency, merely to distinguish bigger urban centers from smaller ones. All-uppercase letters were used for state capitals (such as Albany, New York, and Montpelier, Vermont) as well as for big cities such as New York City.
26. I included Albion, an observer site on Lake Ontario just above the label for Lockport, where the sole X-sign within New York State looks like a plus-sign rotated 45°, and associated "N. Salem" on the map with "Salem Centre" in the list. On another matter, the map's treatment of Albion suggests that place labels were applied to a wax engraving with a stamping stick, used in this case by setting the type for "+ Albion" on a straight stamping stick, and then applying the label to the engraving at an angle of roughly 45°, to fit between the *T* and *R* in "L. ONTARIO." Imprecise stamping might account for some (but surely not all) of the map's locational discrepancies. For information on the use of stamping sticks to apply labels to wax engravings, see David Woodward, *The All-American Map: Wax Engraving and Its Influence on Cartography* (Chicago: University of Chicago Press, 1977), 62–68.
27. *Smithsonian Report*, 1859, 47.

28. *Smithsonian Report*, 1859, 48. Weather observers, as organized by the New York Regents and the Smithsonian Institution, constituted what Jeremy Vetter called a "field network—linking together lay observers in geographically distributed locations with a central figure who aggregated their locally produced observations into more general, regional knowledge." See Jeremy Vetter, "Lay Observers, Telegraph Lines, and Kansas Weather: The Field Network as a Mode of Knowledge Production," *Science in Context* 24 (2011): 259–80, quotation on 259.
29. *Smithsonian Report*, 1957, 26.
30. Not every volunteer received a full set of instruments, consisting of a barometer, hygrometer, rain gauge, and thermometer. Because the Smithsonian's board had appropriated only a thousand dollars, some received only a thermometer, and some received nothing but instructions for noting the general state of weather, including rain, wind, and sky conditions. *Smithsonian Report*, 1848, 15.
31. *Smithsonian Report*, 1857, 28. The computational effort was outsourced to James Coffin, a professor at Lafayette College, in Easton, Pennsylvania. Coffin's clerks worked part time—many were school teachers—and their pay was the equivalent of only three or four full-time workers. Fleming, *Meteorology in America*, 126.
32. In his 1856 report, Henry noted that the new set of blank forms, which he had designed, had been "widely distributed [to observers] under the frank of the Patent Office." *Smithsonian Report*, 1856, 34.
33. *Smithsonian Report*, 1852, 23.
34. For examples, see *Smithsonian Report*, 1852, 81–82; *Smithsonian Report*, 1853, 61–62; and *Smithsonian Report*, 1854, 51.
35. See, for example, *Smithsonian Report*, 1852, 83.
36. *Smithsonian Report*, 1860, 35.
37. Mason, who served from 1853 to 1857 and left to practice law in Iowa, was succeeded by Joseph Holt, who stepped down in 1859 to become postmaster general and was followed by William Darius Bishop, who left after less than a year in office to become secretary of the Treasury. United States Patent and Trademark Office, "Past Leaders of the USPTO," accessed March 10, 2018, https://www.uspto.gov/about-us/past-uspto-leaders.
38. *Smithsonian Report*, 1859, 43.
39. *Smithsonian Report*, 1860, 34–35, quotation on 35.
40. These informal weather reports sometimes warned of approaching storms that could disrupt service: useful information insofar as severe winds could break connections, and excess moisture on the wire could interfere with telegraph signals. Because knowing about a possible disruption was useful in prioritizing transmissions, these reports had become a "routine habit, if not a duty, of many operators." Fleming, *Meteorology in America*, 142–43.
41. *Smithsonian Report*, 1858, 52.
42. *Smithsonian Report*, 1859, 273.
43. *Smithsonian Report*, 1859, 274.
44. *Smithsonian Report*, 1869, 50; and Fleming, *Meteorology in America*, 143, 145.
45. Fleming, *Meteorology in America*, 143.

46. Smithsonian Institution Archives, "Louise Hoover's Painting of Joseph Henry," https://siarchives.si.edu/collections/siris_sic_8265.
47. Fleming, *Meteorology in America*, 147–48.
48. Chas. B. Barr, *Telegraph Stations in the United States, the Canadas & Nova Scotia*, scale ca. 1:4,200,000 (Pittsburgh, PA: Wegner & Buechner, 1853). Retrieved from the Library of Congress, accessed March 12, 2018, https://www.loc.gov/item/97683602/.
49. *Smithsonian Report*, 1857, 26.
50. *Smithsonian Report*, 1858, 52.
51. In the 1858 report, Henry acknowledged "the Morse telegraph line for the gratuitous reports which its operators have daily furnished," but by then the name Morse had become meaningless because of numerous mergers and acquisitions. *Smithsonian Report*, 1858, 52. The 1859 and 1860 reports do not mention a company or route. *Smithsonian Report*, 1859, 95; *Smithsonian Report*, 1860, 36. The pattern of dots in figure 5.5 describing the Smithsonian network in 1860 is largely similar, but not identical, to the pattern on Fleming's dot map for the same year. See Fleming, *Meteorology in America*, 145.
52. *Smithsonian Report*, 1860, 36.
53. *Smithsonian Report*, 1861, 35.
54. *Smithsonian Report*, 1861, 35.
55. Marcus Benjamin, "Meteorology," in *The Smithsonian Institution: 1846–1896: The History of Its First Half Century*, ed. George Brown Goode (Washington, DC, 1897), 647–678, esp. 672–73.
56. *Smithsonian Report*, 1862, 29.
57. Benjamin, "Meteorology," 673, 675.
58. Benjamin, "Meteorology," 676; and Fleming, *Meteorology in America*, 153–56.
59. Eric R. Miller, "New Light on the Beginning of the Weather Bureau from the Papers of Increase A. Lapham," *Monthly Weather Review* 59 (1931): 65–70, esp. 68; and Donald R. Whitnah, *A History of the United States Weather Bureau* (Urbana: University of Illinois Press, 1965), 19–20.
60. James Rodger Fleming, "Storms, Strikes, and Surveillance: The US Army Signal Office, 1861–1891," *Historical Studies in the Physical and Biological Sciences* 30, no. 2 (2000): 315–32; James R. Fleming, "Telegraphing the Weather: Military Meteorology, Strategy, and 'Homeland Security' on the American Frontier in the 1870s," in *Instrumental in War: Science, Research, and Instruments between Knowledge and the World*, ed. Steven A. Walton (Boston: Brill Academic Publishers, 2005), 153–78; and Albert J. Myer, *A Manual of Signals: For the Use of Signal Officers in the Field* (Washington, DC, 1864).
61. Cleveland Abbe, "Historical Notes on the Systems of Weather Telegraphy and Especially Their Development in the United States," *American Journal of Science and the Arts*, series 3, 2 (1871): 81–88; Mark W. Harrington, "History of the Weather Map," in *Report of the International Meteorological Congress Held at Chicago, Ill., August 21–24, 1893* (Washington, DC: Weather Bureau, 1896), 327–36, esp. 330; Miller, "Evolution of Meteorological Institutions," 4; and Edmund P. Willis and William H. Hooke, "Cleveland

Abbe and American Meteorology, 1871–1901," *Bulletin of the American Meteorological Society* 87 (2006): 315–26.

62. For an obituary, see Alfred Judson Henry, "Memoir of Cleveland Abbe," *Annals of the Association of American Geographers* 7 (1917): 61–67. Also see Willis and Hooke, "Cleveland Abbe and American Meteorology."
63. "Cipher Used for the Telegraphic Transmission of the Weather-Reports of the Signal Service, United States Army, Prepared under the Direction of the Chief Signal-Officer of the Army," Annual Report of the Chief Signal-Officer to the Secretary of War for the Year 1872 [hereafter *CSO, 1872*], 169–80 [Paper C].
64. Although public administration analyst Gustavus Weber implied a greater efficiency when he wrote, "By this means, an entire observation can be expressed by a telegraphic cipher message of four or five words, which if expanded into descriptive language would require from thirty to forty words," use of only numbers and abbreviations would have afforded significant compression. Use of familiar words found in a dictionary avoided transcription, coding, and decoding errors more likely if numbers were used. Gustavus A. Weber, *The Weather Bureau: its History, Activities and Organization*, Institute for Government Research monograph no. 9 (New York: D. Appleton and Co., 1922), 18.
65. The Signal Service's annual reports described the circuits collectively in a single table listing their length and membership and individually in twenty-five tables focused on their "working forms." *CSO, 1871*, 57, 115–33.
66. Weber, *The Weather Bureau*, 8–9.
67. At the bottom of each schedule were four notes, the last of which read, "The time given for sending reports is the local time of that particular station." See, for example, *CSO, 1871*, 117.
68. *CSO, 1871*, 57, 117.
69. Present-day longitudes of 71.06° W and 73.21° W for Boston and Burlington, respectively, yield a difference of 2.15°, which represents a time difference of 8.6 minutes.
70. *CSO, 1871*, 58.
71. The schedule for the Milwaukee–New York circuit lists 28 specific cities, including San Francisco and New Orleans, for which data were placed on the Milwaukee–New York circuit. *CSO, 1871*, 115–33, esp. 122.
72. *CSO, 1871*, 34, 116.
73. *CSO, 1871*, 129, 132–33. Circuit schedules do not mention any electronic flow of weather data beyond Chicago.
74. Fleming, "Telegraphing the Weather," 165–67.
75. *CSO, 1872*, 140.
76. *CSO, 1872*, 140.
77. Harrington, "History of the Weather Map," table II, on 336, lists seventy-three stations issuing weather maps on June 1, 1893.
78. *CSO, 1872*, 147–48, 150–51.
79. "Instructions for Stamping Manifold Weather-Maps," *CSO, 1872*, 148–49, quotations on 148.
80. *CSO, 1872*, 141–42.

81. Quotation from Instruction 19. *CSO, 1872*, 142.
82. W. B. Hazen, *History of the Signal Service, with Catalogue of Publications, Instruments and Stations [Prepared under the direction of Brig. & Bvt. Maj. Gen'l W. B. Hazen, Chief Signal Officer of the Army]* (Washington City, 1884), 21–24.
83. Eviatar Zerubavel, "The Standardization of Time: A Sociological Perspective," *American Journal of Sociology* 88 (1982): 1–23; and Vanessa Ogle, *The Global Transformation of Time, 1870–1950* (Cambridge, MA: Harvard University Press, 2015), 25–26.
84. For various changes between 1870 and 1888 in the nature, timing, and frequency of weather observations, see Weber, *The Weather Bureau*, 7.
85. *CSO, 1871*, 58; *Report of the Chief of the Weather Bureau for 1891* [hereafter *Weather Bureau Report*], 288–89.
86. Weber, *The Weather Bureau*, 18–19, quotation on 18.
87. Weber, *The Weather Bureau*, 18–19. For facts about the reach and cost of the Weather Bureau telegraph network, including a highly favorable cost-benefit estimate, see US Bureau of the Census, *Special Reports: Telephones and Telegraphs, 1902* (Washington, DC), 109–10.
88. Edgar B. Calvert, "Development of the Daily Weather Map," *Proceedings of the Convention of Weather Bureau Officials*, Omaha, Nebr*aska*, October 13–14, 1898, US Weather Bureau Bulletin no. 24 (Washington, DC, 1899), 145–46; and Monmonier, *Air Apparent*, 158–60. The map was reproduced in the *New York Daily Graphic*, a pioneering illustrated newspaper, which initiated a daily weather map on May 9, 1879, but discontinued the graphic three years later.
89. Robert DeCourcy Ward, "The Newspaper Weather Maps of the United States," *American Meteorological Journal* 11 (July 1894): 96–107. For insights on Ward's career, including wanderings into eugenics and immigration that tarnished his reputation as a climatologist, see Robert V. Rohli and Gregory D. Bierly, "The Lost Legacy of Robert DeCourcy Ward in American Geographical Climatology," *Progress in Physical Geography* 35 (2011): 547–64.
90. Henry L. Heiskell, "The Commercial Weather Map of the United States Weather Bureau," *Yearbook of the Department of Agriculture: 1912* (Washington, DC), 537–39.
91. Monmonier, *Air Apparent*, 167; and *Weather Bureau Report*, 1912, 24.
92. "AP Wirephotos Flash Across Nation," *Editor and Publisher* 67 (5 January 1935): 7; John W. Perry, "Wirephoto Found Flexible and Speedy; Operates with Uncanny Precision," *Editor and Publisher* 67 (12 January 1935): III and XI; and Mark Monmonier, "Telegraphy, Iconography, and the Weather Map: Cartographic Weather Reports by the United States Weather Bureau, 1870-1935," *Imago Mundi* 40 (1988): 15–31, esp. 26–28. For a concise history of facsimile image transmission, see Jennifer S. Light, "A Forgotten 'New Medium' from the 20th Century," *New Media & Society* 8 (2006); 355–78.
93. Donald R. Whitnah, *A History of the United States Weather Bureau* (Urbana: University of Illinois Press, 1965), 229–30.
94. George Albert Gladney, "The McPaper Revolution? *USA Today*-Style Innovation at Large US Dailies," *Newspaper Research Journal* (Winter/Spring 1992): 54–71; Monmonier, *Air Apparent*, 170–74; and Monmonier, *Maps with the News: The Development of American Journalistic Cartography* (Chicago: University of Chicago Press, 1989), 118–22.

95. An example is the Pennsylvania State University Meteorology Department's "Forecasters e-Wall," online at http://mp1.met.psu.edu/~fxg1/ewall.html.
96. Mark W. Harrington, "The Weather Map," in *Weather Bureau Report*, 1891, 548–53; and Harrington, "History of the Weather Map," 330–31.
97. George S. Bliss, "The Weather Business: A History of Weather Records and the Work of the US Weather Bureau," *Scientific American Supplement* 84 (1917): 110–11, quotation on 110. For a short biography of Bliss, see NOAA Central Library, NOAA History, *Profiles in Time—NWS Biographies*, http://www.history.noaa.gov/nwsbios/nwsbios_page4.html#g_bliss. Bliss was author of *Weather Forecasting with Introductory Note on Atmospherics*, 2d ed. (US Department of Agriculture Weather Bureau Bulletin no. 42, 1917). Other editions were published in 1913, 1921, 1925, 1929, and 1939.
98. Alfred J. Henry, "Frank Hagar Bigelow (1851–1924)," *Monthly Weather Review* 52 (1924): 165–66.
99. Frank H. Bigelow, *Storms, Storm Tracks, and Weather Forecasting,* US Weather Bureau Bulletin no. 20 (Washington, DC, 1897).
100. Bigelow, *Storms, Storm Tracks, and Weather Forecasting*, 14.
101. The ever-vigilant receiver continually monitors a specific all-hazards channel, listening for a signal matching its programmed location. Should it receive a location-specific warning, the receiver turns up the volume and sounds a tone alarm to get residents' attention. John H. Sorensen, "Hazard Warning Systems: Review of 20 Years of Progress," *Natural Hazards Review* 1, no. 2 (2000): 119–25. Originally, the zones were counties or parts of counties. More recently, NOAA introduced storm-based warning technology intended to reduce false alarms based on overly large county-unit warning areas. See "National Weather Service Storm Based Warnings," https://www.weather.gov/media/pah/WeatherEducation/stormbased.pdf.
102. National headquarters, National Weather Service, "History of the National Weather Service," accessed May 13, 2018, https://www.weather.gov/timeline.
103. Monmonier, *Air Apparent*, 95–97; and Weather Bureau, *Operations of the Weather Bureau*, April 1, 1967 (Silver Spring, MD: US Dept. of Commerce, Environmental Services Administration, Weather Bureau, 1967), esp. 12–22.
104. Committee on the Assessment of the National Weather Service's Modernization Program, Board on Atmospheric Sciences and Climate, National Research Council, *The National Weather Service Modernization and Associated Restructuring: A Retrospective Assessment* (Washington, DC: National Academies Press, 2012), 1–2.
105. For a state-by-state historical account of the openings, closings, and relocations of local weather service offices, see Central Illinois Weather Forecast Office, National Weather Service, "Brief History of National Weather Service Offices Past and Present," https://www.weather.gov/ilx/nws-wb-history.
106. For a concise overview of the development and deployment of WSR-88D system, see Monmonier, *Air Apparent*, 143–52.
107. The Department of Commerce, which includes the National Weather Service, operates 122 NEXRAD sites in the conterminous United States; the Department of Transportation

is responsible for twelve sites in Alaska, Hawaii, and Puerto Rico; and the Department of Defense operates twenty-one sites in the conterminous United States, one in Guam, and three in Hawaii. NOAA's National Weather Service Radar Operations Center, "List of NEXRAD Sites," accessed May 13, 2018, https://www.roc.noaa.gov/WSR88D/Program/NetworkSites.aspx.

108. These example distances were calculated as median maximum ranges, which tend to be somewhat greater than the actual distances experienced by trained forecasters working in the early 1990s. NEXRAD Panel, National Weather Service Modernization Committee, Commission on Engineering and Technical Systems, National Research Council, *Toward a New National Weather Service: Assessment of NEXRAD Coverage and Associated Weather Services* (Washington, DC: National Academy Press, 1995), 16.
109. Terrain or available land typically dictated a physical separation. I estimated straight-line distances using Google Maps and radar coordinates archived at https://catalog.data.gov/dataset/next-generation-radar-nexrad-locations and confirmed locations using the Google Maps "Satellite View" overhead images. Albany's radar is on an elevated site south-southwest of the WFO, on a state university campus, whereas the Binghamton and Buffalo radars are within a mile of the WFO and essentially on airport property, and the Burlington radar is only three miles away, in a secure location on an Army National Guard base—an essential separation because the WFO is in the airport terminal building.
110. Mark Monmonier, *Lake Effect: Tales of Large Lakes, Arctic Winds, and Recurrent Snows* (Chicago: University of Chicago Press, 2012), 124.
111. NOAA's National Weather Service Radar Operations Center, "List of NEXRAD sites."
112. Paths of storms were a key factor in locating NEXRAD stations. See D. A. Leone, R. M. Endlich, J. Petričeks, R. T. H. Collis, and J. R. Porte, "Meteorological Considerations Used in Planning the NEXRAD Network," *Bulletin of the American Meteorological Society* 70 (1989): 4–13.
113. Monmonier, *Lake Effect*, 78–79. For examples of related research publications, see Thomas A. Niziol, "Operational Forecasting of Lake-Effect Snowfall in Western and Central New York," *Weather and Forecasting* 2 (1987): 310–21; and Thomas A. Niziol, Warren R. Snyder, and Jeff S. Waldstreicher, "Winter Weather Forecasting throughout the Eastern United States. Part IV: Lake Effect Snow," *Weather and Forecasting* 10 (1995): 61–77.
114. For a concise critical assessment of ASOS, see Committee on the Assessment of the National Weather Service's Modernization Program, *National Weather Service Modernization and Associated Restructuring*, esp. 47–49.
115. Locations, broadcast frequencies, and phone numbers are available at Federal Aviation Administration, Surface Weather Observation Stations (ASOS/AWOS), https://www.faa.gov/air_traffic/weather/asos/.
116. NOAA National Centers for Environmental Information, "Automated Surface Observing System (ASOS)," accessed May 17, 2018, https://www.ncdc.noaa.gov/data-access/land-based-station-data/land-based-datasets/automated-surface-observing-system-asos.
117. NOAA National Centers for Environmental Information, "Cooperative Observer Network (COOP)," Accessed May 17, 2018, https://www.ncdc.noaa.gov/data-access/

land-based-station-data/land-based-datasets/cooperative-observer-network-coop. Also see Christopher A. Fiebrich, “History of Surface Weather Observations in the United States,” *Earth Science Reviews* 93 (2009): 77–84.

Chapter 6

1. The boundary includes a portion of the Missouri River and a portion of the meridian running south from the early nineteenth-century confluence of the Kansas and Missouri Rivers—although erosion and deposition have moved the junction point, the meridian boundary is fixed at approximately 94.61° W. Despite the border’s long and lingering historical, political, and cultural significance, the longitude of its straight, north-to-south portion is rarely mentioned, and is nowhere to be found in Jeremy Neely, *The Border Between Them: Violence and Reconciliation on the Kansas–Missouri Line* (Columbia, MO: University of Missouri Press, 2007).
2. For a historical overview of the topological structuring of cartographic data, see Donna J. Peuquet, “Electronic Cartography: Data Structures and the Storage and Retrieval of Spatial Data,” *HC6*, 359–66.
3. The first two letters are abbreviations for the soil’s name, and the third letter, capitalized by convention, describes the range of slope as 0 to 3 percent (A) or 3 to 8 percent (B).
4. Compound islands are also known as *complex islands* and *multipart polygons*.
5. USGS land-use/land-cover data are based on level II of a classification system developed for overhead imagery. A land polygon is not included in the data if its area is less than four hectares and it would be classified as urban and built-up; water; confined feeding and “other agricultural land”; or strip mine, quarry, or gravel pit. For other categories, including forest land and wetland, the minimum mapping unit is 16 hectares. See William B. Mitchell, Stephen C. Guptill, K. Eric Anderson, Robin G. Fegeas, and Cheryl A. Hallam, *GIRAS: A Geographic Information Retrieval and Analysis System for Handling Land Use and Land Cover Data*, USGS Professional Paper 1059 (Washington, DC, 1977), 2–3. In GIRAS lingo, a multipart polygon is a *compound island*.
6. Complex islands (multipart polygons) are accorded special treatment, typically with each complex island recognized as a separate entity with its own index number, its own perimeter, its own list of constituent islands, and (instead of a land category attribute) a flag marking it for special treatment. For a discussion of complex islands, see Jan W. van Roessel and Stuart W. Doescher, *Vector Data Conversion at the EROS Data Center*, USGS Open File Report 87-309 (Sioux Falls, SD, 1986), esp. 13–15.
7. Although proprietary illustration and image software such as Adobe Illustrator and Photoshop might be more efficient, the overall process and intended result are similar.
8. This “debate” extended roughly from around 1978 through 1990. Robert B. McMaster, “Electronic Cartography: Intellectual Movements in Electronic Cartography,” *HC6*, 382–85, quotation on 382; and Donna J. Peuquet, “Electronic Cartography: Data Structures,” *HC6*, 364.

9. For a concise explanation of point-in-polygon screening, see Mark Monmonier, *Computer-Assisted Cartography: Principles and Prospects* (Englewood Cliffs, NJ: Prentice-Hall, 1982), 118–19.
10. Peuquet, "Electronic Cartography: Data Structures," 362–63.
11. US Geological Survey, National Map: Elevation Point Query Service, accessed June 11, 2018, https://nationalmap.gov/epqs/. Older nationwide elevation databases afforded an accuracy, at the 95 percent confidence level, of only 3.04 m. US Geological Survey, "What is the vertical accuracy of the seamless 3DEP DEMs?" accessed June 11, 2018, https://www.usgs.gov/faqs/what-vertical-accuracy-seamless-3dep-dems.
12. Samantha T. Arundel, Christy-Ann M. Archuleta, Lori A. Phillips, Brittany L. Roche, and Eric W. Constance, "1-Meter Digital Elevation Model Specification," *US Geological Survey Techniques and Methods*, book 11, chap. B7 (Reston, VA: USGS, 2015), esp. 4. Elevations are measured using a highly accurate aerial survey technology known as *lidar* (light detection and ranging). For a general description and history of lidar, see Jason M. Stoker and Susan K. Greenlee, "Lidar," *HC6*, 774–76. Until the one-meter dataset was complete for the entire conterminous United States (CONUS), the National Elevation Dataset (NED) provided elevation estimates based on digital elevation models (DEMS) with a lower horizontal and vertical resolution.
13. Because a slight overlap can expedite calculations, blocks are distributed as 10,006 × 10,006 DEMs. Arundel and others, "1-Meter Digital Elevation Model Specification," 2–3. Earlier, less exact elevation datasets produced for DEMs with only a one-degree, one-arc-second, two-arc second, or ten-meter resolution were distributed as single blocks for each one-degree tile. Christy-Ann M. Archuleta, Eric W. Constance, Samantha T. Arundel, Amanda J. Lowe, Kimberly S. Mantey, and Lori A. Phillips, "The National Map Seamless Digital Elevation Model Specifications," *US Geological Survey Techniques and Methods*, book 11, chap. B9 (Reston, VA: USGS, 2017), 2.
14. On a conventional printed map, the horizontal accuracy of a position is, at best, 0.02 in.—roughly the thickness of a printed line—and elevations were assumed to be no better than within half a contour interval. Before the use of lidar, flood insurance maps were notoriously inaccurate. Committee on FEMA Flood Maps, National Research Council, *Mapping the Zone: Improving Flood Map Accuracy* (Washington, DC: National Academy Press, 2009), 30, 107.
15. The National Map Accuracy standards are based on a validation test of "well-defined" points, readily recognized on both the map and the landscape. No more than 10 percent of the points tested may differ from the correct value by more than half a contour interval. Committee on Floodplain Mapping Technologies, National Research Council, *Elevation Data for Floodplain Mapping* (Washington, DC: National Academy Press, 2007), 61–63; and US Geological Survey, "Map Accuracy Standards," Fact Sheet 171-99 (November 1999).
16. For a concise discussion of triangulated irregular networks (TINs), including methods for constructing a TIN database, see John P. Wilson, *Environmental Applications of Digital Terrain Modeling* (Hoboken, NJ: Wiley-Blackwell, 2018), 24–29.

17. Contour threading can address special cases of a contour line that enters a triangle at a node as well as the unusual instance of a contour that follows an edge. Rarer still, and probably precluded by the process that selects nodes, is the closed contour that encompasses a triangle's entire perimeter. The triangular facets of a terrain surface almost always have some tilt upward from the horizontal.
18. The gray line in figure 6.5 represents a worked example. A contour line with an elevation of 1,500—feet or meters, take your pick—enters triangle *M* midway between nodes with elevations 1,000 and 2,000. It must then cross into triangle *G*, through an edge bracketed by elevations of 1,000 and 1,667 because both the elevations bracketing the other edge exceed 1,500. Because 1,500 is 75 percent of the difference (667) between 1,000 and 1,667, the contour crosses this edge three quarters of the distance from the node at 1,000. My contrived elevations simplify percentage displacements of the contour crossings, indicated by small open circles. Note that the contour line always crosses an edge with one node higher and the other node lower. You can follow the contour from triangle *G* through triangles *H*, *K*, *J*, and *I*.
19. The example was created in Adobe Illustrator, which uses Bézier spline interpolation to convert a series of linked straight-line segments into a smooth curve. For discussion of Bézier curves, see Sharon Steuer, *The Adobe Illustrator CS Wow! Book* (Berkeley, CA: Peachpit Press, 2004), 6–9.
20. For a fuller discussion of what's called the "hidden-line problem," see James D. Foley and Andries Van Dam, "Algorithms for Removing Hidden Edges and Surfaces," in their *Fundamentals of Interactive Computer Graphics* (Reading, MA: Addison-Wesley, 1982), 553–73; and Monmonier, *Computer-Assisted Cartography*, 134–36.
21. For a discussion of how block data support the crafting of visually complex election districts that serve partisan aims while addressing the US Constitution's concerns about fairness and racial equality, see Mark Monmonier, *Bushmanders and Bullwinkles: How Politicians Manipulate Electronic Maps to Win Elections* (Chicago: University of Chicago Press, 2001), esp. 104–9. Strictly speaking, the constitutional mandate for the US Census is tied to congressional reapportionment, which refers to the number of seats a state has in the House of Representatives: each census reveals population shifts, which often take seats from some states and give them to others. However, increasingly strict court interpretation of the one-person/one-vote principle (the Equal Protection Clause of the Fourteenth Amendment) has been tied to block data, which allow the configuration of congressional districts with little variation in population.
22. For information on the 1970 Census enumeration, see US Bureau of the Census, *1970 Census Users' Guide*, part I (Washington, DC: US Government Printing Office, 1970), esp. 23–28.
23. G. Harry Stine helped popularize the term "Third Industrial Revolution" in his so-named *The Third Industrial Revolution* (New York: Ace Books, 1979). For a more recent overview of the environmental impacts, see Jeremy Rifkin, *The Third Industrial Revolution: How Lateral Power Is Inspiring a Generation and Transforming the World* (New York: Palgrave Macmillan, 2011).

24. Kenneth J. Dueker, "Geocoding," *HC6*, 448–52, esp. 449; and Mark Monmonier and Robert W. Marx, "Census Mapping," *HC6*, 207–12, esp. 209.
25. Gavin H. Shaw, CREATE, *A Supervisor's Manual: Supervisory Procedures for the Creation or Extension of a GBF/DIME System* (Washington, DC: Bureau of the Census, 1976), 30.
26. "Middle forty-eight" replaced "lower forty-eight" when Hawaii, which is farther south, became the fiftieth state in 1959, several months after Alaska was admitted as the forty-ninth state. *Conterminous*, the official term, has the same meaning as *contiguous*.
27. For an obituary, see Marvin White, "In Memoriam: James P. Corbett (1909–1998)," *Cartography and Geographic Information Systems* 25 (1998): 264–65.
28. James P. Corbett, *Topological Principles in Cartography*, Technical Paper no. 48 (Washington, DC: US Bureau of the Census, 1979).
29. George Farnsworth, *Census Use Study: DIME, a Geographic Base File Package* (Washington, DC: Bureau of the Census, 1972), 38.
30. Farnsworth, *Census Use Study: DIME*, 38–40.
31. Quotation from Monmonier, *Computer-Assisted Cartography*, 151, which is based on James P. Corbett, "Topological Principles in Cartography," in *Proceedings of the International Symposium on Computer-Assisted Cartography, Auto-Carto II*, September 21–25, 1975 (Washington, DC: US Bureau of the Census; Falls Church, VA: American Congress on Surveying and Mapping, 1977), 61–65; and Marvin S. White Jr., "A Geometrical Model for Error Detection and Correction," in *Proceedings of the International Symposium on Computer-Assisted Cartography, Auto-Carto III*, January 16–20, 1978 (Falls Church, VA: American Congress on Surveying and Mapping, 1979), 439–56, esp. 444.
32. White, "A Geometrical Model for Error Detection and Correction," 451–52.
33. E. W. Dijkstra, "A Note on Two Problems in Connexion with Graphs," *Numerische Mathematic* 1 (1959): 269–71. For enhancements to Dijkstra's technique and comparisons with competing methods, see L. Fu, D. Sun, and L. R. Rilett, "Heuristic Shortest Path Algorithms for Transportation Applications: State of the Art," *Computers and Operations Research* 33 (2006): 3325–43; Stefanie Kritzinger, Karl F. Doerner, Richard F. Hartl, Günter Kiechle, Horst Stadler, and Senthanal Sirpi Manohar, "Using Traffic Information for Time-Dependent Vehicle Routing," *Procedia—Social and Behavioral Sciences* 39 (2012): 217–29; and Detian Zhang, Yuan Liu, An Liu, Xudong Mao, and Qing Li, "Efficient Path Query Processing through Cloud-Based Mapping Services," *IEEE Access* 5 (2017): 12963–73.
34. John Markoff, "Edsger Dijkstra, 72, Physicist Who Shaped Computer Era," *New York Times*, August 10, 2002, A11.
35. David M. Mark, Michael D. Gould, and Matthew McGranaghan, "Computerized Navigation Assistance for Drivers," *Professional Geographer* 39 (1987): 215–20.
36. For discussion of time-dependent routing issues, see Kritzinger and others, "Using Traffic Information for Time-Dependent Vehicle Routing," *Procedia - Social and Behavioral Sciences* 39 (2012): 217 – 229.
37. Winnie Hu, "Is Your E-Z Pass the Key to Congestion Pricing?" *New York Times*, March 21, 2018.

38. Gary Burnett, "'Turn Right at the Traffic Lights': The Requirement for Landmarks in Vehicle Navigation Systems," *Journal of Navigation* 53 (2000): 499–510.
39. For examples of the discussion surrounding map-matching algorithms, including concerns for computational efficiency—important because these calculations are so frequent—see Bi Yu Chen, Hui Yuan, Qingquan Li, William H. K. Lam, Shih-Lung Shaw, and Ke Yan, "Map-Matching Algorithm for Large-Scale Low-Frequency Floating Car Data," *International Journal of Geographical Information Science* 28 (2014): 22–38; Mahdi Hashemi and Hassan A. Karimi, "A Critical Review of Real-Time Map-Matching Algorithms: Current Issues and Future Directions," *Computers, Environment, and Urban Systems* 48 (2014): 153–65; Mohammed A. Quddus, Robert B. Noland, and Washington Y. Ochieng, "Validation of Map-Matching Algorithms Using High Precision Positioning with GPS," *Journal of Navigation* 58 (May 2005): 257–71; and Sébastien Romon, Xavier Bressaud, Sylvain Lassarre, Guillaume Saint Pierre, and Louahdi Khoudour, "Map-Matching Algorithm for Large Databases," *Journal of Navigation* 68 (2015): 971–88. Inertial navigation technology further increases accuracy by providing a reliable means of interpolating between GPS estimates of a moving vehicle's location. Inertial navigation was especially important in the 1990s, when the GPS time signal was deliberately blurred because of national security concerns. See J. Mitchell, "The Application of Inertial Navigation Technology in Land Vehicles," *Journal of Navigation* 48 (January 1995): 81–87.
40. For information on tiling used by web map services such as Google Maps, see Ian Muehlenhaus, *Web Cartography: Map Design for Interactive and Mobile Devices* (Boca Raton, FL: CRC Press, 2014), 208–16; Felix Ortag, Manuela Schmidt, and Georg F. Gartner, "Web-based Wayfinding," *HC6*, 1739–42, esp. 1740; and Michael P. Peterson, *Mapping in the Cloud* (New York: Guilford Press, 2014), 29–33, 370–72.
41. For information about the TMC, see Emere Arco, Andrea Ajmar, Fabrizio Arneodo, and Piero Boccardo, "An Operational Framework to Integrate Traffic Messaging Channel (TMC) in Emergency Mapping Services (EMS)," *European Journal of Remote Sensing* 50 (2017): 478–95; and Marc Heddebaut, Jean-Pierre Ghys, Michael Sanz, and Fouzia Elbahhar, "Road Traffic Information Using a Dedicated Radio Beacon," *Journal of Transportation Research—Part C, Emerging Technologies* 35 (2013): 20–33.
42. Google's "official blog" described the service, the technology, and the company's intentions. For an example, see Dave Barth, "The Bright Side of Sitting in Traffic: Crowdsourcing Road Congestion Data," August 25, 2009, *Google Official Blog*. Barth was product manager for Google Maps.
43. For an overview of the ethics and reliability of crowdsourced data, see Sarah Elwood, Michael F. Goodchild, and Daniel Z. Sui, "Researching Volunteered Geographic Information: Spatial Data, Geographic Research, and New Social Practice," *Annals of the Association of American Geographers* 102 (2012): 571–90.
44. See, for example, Mohammad Reza Jabbarpour, Ali Jalooli, Erfan Shaghaghi, Rafidah Md Noor, Leon Rothkrantz, Rashid Hafeez Khokhar, and Nor Badrul Anuar, "Ant-Based Vehicle Congestion Avoidance System Using Vehicular Networks," *Engineering Applications of Artificial Intelligence* 36 (2014): 303–19.

45. Lisa W. Foderaro, "Navigation Apps Are Turning Quiet Neighborhoods into Traffic Nightmares," *New York Times*, December 24, 2017, https://nyti.ms/2pnWkOB; and John Surico, "What Happens When a City Bans Non-Resident Drivers?" CityLab, April 18, 2018, https://www.citylab.com/transportation/2018/04/the-small-town-that-took-on-waze/558215/.
46. S. P. Sullivan, "No, You Can't Ban Out-of-Towners from Your Roads, A.G. Tells Traffic-Tied Towns," NJ Advance Media for NJ.com, May 3, 2018, https://www.nj.com/politics/index.ssf/2018/05/out-of-town_traffic_bans_are_illegal_ag_tells_traf.html.
47. See, for example, Eileen Sullivan, "Law Enforcement Wants Police-Tracking App Disabled," *Syracuse Post-Standard*, January 15, 2015, B-3. Google purchased Waze in 2013. For a technical description of most of Waze's functionality, see Uri Levine, Amir Shinar, and Ehud Shabtai, "System and Method for Road Map Creation," US Patent 8,612,136, filed August 27, 2008, and issued December 17, 2013. The patent had been assigned to WAZE Mobile Ltd., an Israeli firm.
48. Aarian Marshall, "Waze Digs into Your Car's Dashboard," Wired.com, February 27, 2017, https://www.wired.com/2017/02/waze-cars/.
49. A GPS vehicle navigation system patented in 2016 specifically mentioned red-light cameras. William Carl Stumphauzer and Todd Darrell Van Zandt, "System and Method for Improved Updating and Annunciation of Traffic Enforcement Camera Information in a Vehicle Using a Broadcast Content Delivery System," US Patent 9,466,212, filed January 5, 2011, and issued October 11, 2016.
50. Roberto Carisi, Eugenio Giordano, and Giovanni Pau, "Enhancing In-Vehicle Digital Maps via GPS Crowdsourcing," *UCLA Computer Science Department Technical Report* #TR100016, 2011; and Tom-Tom N.V., "Historical Travel Information: White Paper," geospatial.tomtom.com, accessed February 12, 2019, https://www.tomtom.com/lib/img/HISTORICAL_TRAFFIC_WHITEPAPER.pdf.
51. For discussion of path caching, see Detian Zhang, Yuan Liu, An Liu, Xudong Mao, and Qing Li, "Efficient Path Query Processing through Cloud-Based Mapping Services," *IEEE Access* 5 (2017): 12963–73.
52. Reza Sarraf and Michael D. McGuire, "A Data Driven Approach for Safe Route Planning," *International Journal of Applied Geospatial Research* 9 (2018): 1–18.
53. Theo Arentze, Tao Feng, Jops Robroeks, Marc van Brakel, and Rob Huibers, "Compliance with and Influence of a New In-Car Navigation System for Trucks: Results of a Field Test," *Transport Policy* 23 (2012): 42–49.
54. Marnie Eisenstadt, "After Years of Anguish, Driver in Megabus Crash Dies at 63," *Syracuse Post-Standard*, March 17, 2015, A-1.
55. For an overview of PTC, see Alan Polivka, Bill Moore Ede, and Joe Drapa, *North American Joint Positive Train Control Project*, Report DOT/FRA/ORD-09/04 (Washington, DC: Federal Railroad Administration, 2009); Sharon Niel, ed., *Rail Safety: Challenges, Oversight Issues, and Positive Train Control* (Hauppauge, NY: Novinka, 2014); John DeWaele, "Positive Train Control: Panacea or Partial Solution," *Railpace Newsmagazine* 37, no. 4 (April 2018): 38, 43; and Brian Solomon, "Willpower, Teamwork & Good Luck:

SEPTA's PTC System Works in Philadelphia, Years Before Others," *Trains* 78, no. 7 (July 2018): 24–33.

56. Matthew Shaer, "The Wreck of Amtrak 188," *New York Times Magazine*, January 31, 2016.
57. National Transportation Safety Board, "Derailment of Amtrak Passenger Train 188, Philadelphia, PA, May 12, 2015," *Public Meeting of May 17, 2016*, Report NTSB/DCA15MR010, p. 1 of Executive Summary, accessed July 4, 2018, http://www.ntsb.gov/news/events/Documents/DCA15MR010_Abstract.pdf.
58. Polivka and others, *North American Joint Positive Train Control Project*, 7, 51–52; and Solomon, "Willpower, Teamwork & Good Luck." Precise location and train speed are also provided by high-frequency, low-voltage data pulsed through the rails.
59. Eric Jaffe, "The Billion-Dollar Technology that May or May Not Prevent the Next Big Train Crash," CityLab, July 31, 2013; Ashley Halsey, III, and Michael Laris, "Deadline for Train Safety Technology Undercut by Industry Lobbying," *Washington Post*, October 25, 2015; and Ashley Halsey III and Michael Laris, "Senate Passes Transportation Funding Stopgap Bill and Rail-Safety Extension," *Washington Post*, October 28, 2015. The statute allowed a further extension to December 31, 2020, for railroads that were making substantial progress toward full compliance. Railroad companies are required to file a quarterly compliance report with the Federal Railroad Administration, which posted a summary suggesting that as of December 31, 2018, full compliance was in sight for most, but not all, operators. For a press release, see https://railroads.dot.gov/newsroom/fra-publishes-railroads-2018-quarter-4-ptc-data. For a "dashboard" comparing various operating companies, see https://www.fra.dot.gov/app/ptc.

Chapter 7

1. Hod Lipson and Melba Kurman, *Driverless: Intelligent Cars and the Road Ahead* (Cambridge, MA: MIT Press, 2017), 127–58.
2. When a deliberate degradation called *selective availability* (SA) was turned off in 2000, 95 percent of the time uncorrected GPS positions were horizontally accurate to within about 15 m. Map-matching enhanced by inertial navigation technology and augmentation systems such as WAAS reduced this distance to about 3 m. Lawrence Letham, *GPS Made Easy: Using Global Positioning Systems in the Outdoors*, 5th ed. (Seattle: Mountaineers Books, 2003), 23–25; Val Noronha, "Wayfinding and Travel Maps: In-Vehicle Navigation Systems," *HC6*, 1716–22; and Mohammed A Quddus, Washington Y. Ochieng, and Robert B. Noland, "Current Map-Matching Algorithms for Transport Applications: State-of-the Art and Future Research Directions," *Transportation Research Part C: Emerging Technologies* 15, no. 5 (2007): 312–28.
3. Thomas Dautermann, "Civil Air Navigation Using GNSS Enhanced by Wide Area Satellite Based Augmentation Systems," *Progress in Aerospace Sciences* 67 (2014): 51–62; Euiho Kim, Todd Walter, and J. David Powell, "Wide Area Augmentation System-Based Flight Inspection System," *Journal of Aircraft* 45 (March–April 2008): 614–21; Roberto Sabatini, Terry Moore, and Subramanian Ramasamy, "Global Navigation Satellite Systems

Performance Analysis and Augmentation Strategies in Aviation," *Progress in Aerospace Sciences* 95 (2017): 45–98; and Haitao Xiang and Lei Tian, "Method for Automatic Georeferencing Aerial Remote Sensing (RS) Images from an Unmanned Aerial Vehicle (UAV) Platform," *Biosystems Engineering* 108 (2011): 104–13.

4. Edmond Awad, Sohan Dsouza, Richard Kim, Jonathan Schulz, Joseph Henrich, Azim Shariff, Jean-François Bonnefon, and Iyad Rahwan, "The Moral Machine Experiment," *Nature* 563, no. 7729 (2018): 59–64; and Peter Dizikes, "A Self-Driving Car Can Choose Who Dies in a Fatal Crash," *MIT News*, October 29, 2018, accessed February 14, 2019, https://www.weforum.org/agenda/2018/10/how-should-autonomous-vehicles-be-programmed/.
5. Original proposals called for truck platoons with a driver in the lead vehicle and adaptive cruise control maintaining a uniform distance between vehicles. See Anirudh Kishore Bhoopalam, Niels Agatz, and Rob Zuidwijk, "Planning of Truck Platoons: A Literature Review and Directions for Future Research," *Transportation Research Part B: Methodological* 107 (January 2018): 212–28.
6. Meng Lu, Kees Wevers, and Rob Van Der Heijden, "Technical Feasibility of Advanced Driver Assistance Systems (ADAS) for Road Traffic Safety," *Transportation Planning and Technology* 28 (2005): 167–87; and Ankit Gupta and Bhupendra Singh, "Recent Trends in Intelligent Transportation Systems: A Review," *Journal of Transport Literature* 9 (April 2015): 30–34.
7. Lilia Filipova-Neumann and Peter Welzel, "Reducing Asymmetric Information in Insurance Markets: Cars with Black Boxes," *Telematics and Informatics* 27 (2010): 394–403; Tobias Hoppe, Stefan Kiltz, and Jana Dittmann, "Automotive IT-Security as a Challenge: Basic Attacks from the Black Box Perspective on the Example of Privacy Threats," *Proceedings of the International Conference on Computer Safety, Reliability, and Security (SAFECOMP 2009)*, 145–58; and Muhammad Usman Iqbal and Samsung Lim, "Privacy Implications of Automated GPS Tracking and Profiling," *IEEE Technology and Society Magazine* 29, no. 2 (2010): 39–46.
8. Electronic Privacy Information Center, "Automobile Event Data Recorders (Black Boxes) and Privacy," accessed September 24, 2018, https://epic.org/privacy/edrs/.
9. M. S. Kovacevic, K. Gavin, I. Stipanovic Oslakovic, and M. Bacic, "A New Methodology for Assessment of Railway Infrastructure Condition," *Transportation Research Procedia* 14 (2016): 1930–39; David Z. Morris, "Why BNSF Railway Is Using Drones to Inspect Thousands of Miles of Rail Lines," Fortune.com (2015), http://fortune.com/2015/05/29/bnsf-drone-program/; and Graham Warwick, "BNSF Railway Pioneers True Beyond Visual Line of Sight UAS Operations," *Aviation Week & Space Technology*, June 27, 2017, accessed October 15, 2018, http://aviationweek.com/technology/bnsf-railway-pioneers-true-beyond-visual-line-sight-uas-operations.
10. For an example of the hazards of using manned helicopters, see Suzanne Moore and Bob Bennett, "Massena, NJ, Men Killed in Helicopter Crash," *Plattsburg (NY) Press-Republican*, October 31, 2018.

11. For an overview of law enforcement applications, see Philip J. Hiltner, "The Drones Are Coming: Use of Unmanned Aerial Vehicles for Police Surveillance and Its Fourth Amendment Implications," *Wake Forest Journal of Law and Policy* 3, no. 2 (2013): 397–415. For a more apprehensive take on civil liberties issues posed by police drones, see Tyler Wall, "Ordinary Emergency: Drones, Police, and Geographies of Legal Terror," *Antipode* 48, no. 4 (2016): 1122–39. Although little has been written about highway maintenance usage of drones in the United States, Brazilian researchers have addressed applications focused on the condition of road pavement; see Marcos R. Resende, Liedi L. B. Bernucci, and José A. Quintanilha, "Monitoring the Condition of Roads Pavement Surfaces: Proposal of Methodology Using Hyperspectral Images," *Journal of Transport Literature* 8, no. 2 (2014): 201–20.
12. Troy A. Rule, "Drone Zoning," *North Carolina Law Review* 95, no. 1 (2016): 133–200.
13. Mitch Waldrop, "DARPA and the Internet Revolution," in *US Defense Advanced Research Projects Agency, DARPA: 50 Years of Bridging the Gap* (Tampa, FL: Faircount LLC, 2008), 78–85.
14. Waldrop, "DARPA and the Internet Revolution," 80–81.
15. Sean P. Gorman and Edward J. Malecki. "The Networks of the Internet: An Analysis of Provider Networks in the USA," *Telecommunications Policy* 24, no. 2 (2000): 113–34; Barry M. Leiner, Vinton G. Cerf, David D. Clark, Robert E. Kahn, Leonard Kleinrock, Daniel C. Lynch, Jon Postel, Larry G. Roberts, and Stephen Wolff, "A Brief History of the Internet." *ACM SIGCOMM Computer Communication Review* 39, no. 5 (2009): 22–31; and Edward J. Malecki, "The Economic Geography of the Internet's Infrastructure," *Economic Geography* 78, no. 4 (2002): 399–424.
16. Scott Ruthfield, "The Internet's History and Development: From Wartime Tool to Fish-Cam," *Crossroads* 2, no. 1 (1995): 2–4.
17. Douglas Comer, "The Computer Science Research Network CSNET: A History and Status Report," *Communications of the ACM* 26, no. 10 (1983): 747–53; and Peter J. Denning, Anthony Hearn, and C. William Kern, "History and Overview of CSNET," *ACM SIGCOMM Computer Communication Review* 13, no. 2 (1983): 138–45.
18. Bilal Chinoy and Hans-Werner Braun, *The National Science Foundation Network*, NSF Technical Report GA-A21029, SDSC [San Diego Supercomputer Center], 1992.
19. National Science Foundation, Where Discoveries Begin, "NSF and the Birth of the Internet," accessed October 4, 2018, https://www.nsf.gov/news/special_reports/nsf-net/60s_to_90s_map.pdf.
20. Arun Rai, Thiagarajan Ravichandran, and Subhashish Samaddar, "How to Anticipate the Internet's Global Diffusion," *Communications of the ACM* 41, no. 10 (1998): 97–106.
21. For discussion and examples, see Hal Burch and Bill Cheswick, "Mapping the Internet," *Computer* 32, no. 4 (1999): 97–98; Martin Dodge and Rob Kitchen, *Atlas of Cyberspace* (Harlow, England: Addison-Wesley, 2001); Yong Wang, Phillip Lai, and Daniel Sui, "Mapping the Internet Using GIS: The Death of Distance Hypothesis Revisited," *Journal of Geographical Systems* 5, no. 4 (2003): 381–405; D. R. Fraser Taylor, "The Concept of Cybercartography," in *Maps and the Internet*, ed. Michael P. Petersen (Amsterdam: Elsevier,

2005), 405–20; Bradley Fidler and Morgan Currie, "The Production and Interpretation of ARPANET Maps," *IEEE Annals of the History of Computing* 37, no. 1 (2015): 44–55; and Aharon Kellerman, *Geographic Interpretations of the Internet* (Cham, Switzerland: Springer, 2016), esp. 114–19.

22. Quotation from caption for "2007 Internet Map," online at National Science Foundation, Where Discoveries Begin, "NSF and the Birth of the Internet," accessed October 2, 2018, https://www.nsf.gov/news/special_reports/nsf-net/textonly/maps.jsp.
23. Mercedes Bunz and Graham Meikle in their *The Internet of Things* (Cambridge, UK: Polity Press, 2018) provide an accessible overview, reinforced with chapters aptly titled "Addressing Things," "Speaking Things," "Seeing Things," and "Tracking Things."
24. For information on Project SIGMA, which is not an acronym, see "DARPA Seeks to Expand Real-Time Radiological Threat Detection to Include Other Dangers," *DARPA News and Events*, February 20, 2018, accessed October 9, 2018, Outreach@DARPA.mil; "Ushering in a New Generation of Low-Cost, Networked, Nuclear-Radiation Detectors," *DARPA News and Events*, August 23, 2016, accessed October 9, 2018, Outreach@DARPA.mil; Jeffrey L. Lacy, Christopher S. Martin, Athanasios Athanasiades, Murari Regmi, Gerson J. Vazquez-Flores, Stephen Davenport, Nicholas S. King, and Tom Lyons, "Large-Area, Low-Cost, High-Efficiency Neutron Detector for Vehicle-Mounted Operation," *IEEE Transactions on Nuclear Science* 64 (July 2017): 1696–1703.
25. Quotation from video accompanying "Advanced Radioactive Threat Detection System Completes First Large-Scale Citywide Test," *DARPA News and Events*, November 10, 2016, accessed October 9, 2018, Outreach@DARPA.mil.
26. Eric J. Butterbaugh, DARPA Public Affairs, email communication, October 11, 2018.
27. For the official tally of the popular vote (65,853,514 for Clinton, 62,984,828 for Trump), see Federal Election Commission, *Federal Elections 2016: Election Results for the US President, the US Senate and the US House of Representatives* (Washington, DC, 2017), 5. For discussion of disparities between the popular vote and the Electoral College vote, see Drew DeSilver, "Trump's Victory Another Example of How Electoral College Wins Are Bigger than Popular Vote Ones," Pew Research Center, December 20, 2016, accessed October 25, 2018, http://www.pewresearch.org/fact-tank/2016/12/20/why-electoral-college-landslides-are-easier-to-win-than-popular-vote-ones/. For a concise summary of the diverse factors behind Trump's victory, social media excepted, see Henry C. Kenski and Kate M. Kenski, "Explaining the Vote in the Election of 2016: The Remarkable Come from Behind Victory of Republican Candidate Donald Trump," in *The 2016 US Presidential Campaign: Political Campaigning and Communication*, ed. Robert E. Denton Jr. (Cham, Switzerland: Palgrave Macmillan, 2017), 285–309, esp. 305–6. For discussion of Russian hacking of social media, see Alicia Parlapiano and Jasmine C. Lee, "The Propaganda Tools Used by Russians to Influence the 2016 Election," *New York Times*, February 16, 2018; Molly McKew, "Did Russia Affect the 2016 Election? It's Now Undeniable," *Wired*, February 16, 2018, accessed October 25, 2018, https://www.wired.com/story/did-russia-affect-the-2016-election-its-now-undeniable/; and Kathleen Hall Jamieson, *Cyberwar: How*

Russian Hackers and Trolls Helped Elect a President: What We Don't, Can't, and Do Know (New York: Oxford University Press, 2018).

28. Although there are 538 electoral votes, these numbers add up to only 531 because electors who are pledged to a candidate don't always honor their commitment. Defections by so-called "faithless" electors are apparent on the official 2016 map in Hawaii, Texas, and Washington. One Hawaiian elector pledged to Clinton voted for Bernie Sanders, a pair of Texas electors pledged to Trump defected to John Kasich and Ron Paul, and four electors in Washington deserted Clinton: three for Colin Powell and one for Native American activist Faith Spotted Eagle. *Federal Elections, 2016*, 6.
29. Robert C. Turner, "The Contemporary Presidency: Do Nebraska and Maine Have the Right Idea? The Political and Partisan Implications of the District System," *Presidential Studies Quarterly* 35 (March 2005): 116–37. For criticism of the Congressional District system, see Anthony J. McGann, Charles Anthony Smith, Michael Latner, and Alex Keena, "Gerrymandering the Presidency: Why Trump Could Lose the Popular Vote in 2020 by 6 Percent and Still Win a Second Term," *London School of Economics Blog*, February 8, 2017, accessed October 25, 2018, http://bit.ly/2kN9LUG.
30. For a similar map, based on early, unofficial results, see "If All States Voted Like Maine and Nebraska: Trump 290, Clinton 248," 270ToWin, January 5, 2017, accessed October 25, 2018, www.270towin.com/news/.
31. Trump supporters might perceive this as whining, which is fine: I'm an academic, and we do a lot of whining. For anyone who might be interested, I am not registered with any party. I was a Democrat once, and then a Republican, and for the last decade or so, I've been an Independent.
32. Mark Monmonier, *Bushmanders and Bullwinkles: How Politicians Manipulate Electronic Maps and Census Data to Win Elections* (Chicago, University of Chicago Press, 2001), 23–25.
33. Monmonier, *Bushmanders and Bullwinkles*, 2–12, 31–62, quotation on 11. For discussion of the evolution of block-level counts, see US Bureau of the Census, "Census Blocks and Block Groups" (chap. 11) and "Voting Districts" (chap. 14) in *Geographic Areas Reference Manual* (Washington, DC: US Dept. of Commerce, Economics and Statistics Administration, Bureau of the Census, 1994), esp. pages 11-1, 11-3, and 14-10.
34. Quotations from Raymond Hernandez, "Justice Dept. Gives Consent to New Lines for a District," *New York Times*, October 11, 1997; Adam Clymer, "Republicans Gain as States Battle over Redistricting," *New York Times*, December 30, 1991.
35. The pivotal decision was Shaw v. Reno, 509 US 630 (1993). For discussion of the subsequent wrangling that made race acceptable when intertwined with partisan considerations, see Monmonier, *Bushmanders and Bullwinkles*, 42–63.
36. For insights into the role of consultants, who work for only one party, see Michael Wines, "Just How Bad Is Partisan Gerrymandering? Ask the Mapmakers," *New York Times*, January 29, 2018.
37. Luis Gutiérrez represented Illinois' Fourth District from 1993 through 2019, when he retired, and Cardiss Hortense Collins represented the intervening Illinois Seventh District

from 1973 to 1997, when she retired and was succeeded by Danny K. Davis, who remained a successful incumbent after the 2018 elections. Jonathon Berlin and Phil Geib, "Timeline of US Rep. Luis Gutierrez and C-Shaped 4th District," *Chicago Tribune*, November 28, 2017; and Caitlin Mullen, "Davis in Control in 7th District Dem Race, but OPRF Teacher Clark Says He 'Sent the Message,' " *Chicago Tribune*, March 20, 2018.

38. Monmonier, *Bushmanders and Bullwinkles*, 62–63. The Bullwinkle District's representative, Nydia M. Velazquez, first elected in 1993, remained a successful incumbent through the 2018 election; a modified version of her district was renumbered as New York's Seventh District in 2013. Johnathan P. Hicks, "Albany Lawmakers Agree on Redrawn 12th District," *New York Times*, July 31, 1997; and "Rep. Velazquez Posts Congressional Biography on Capitol Hill Web Site," *Targeted News Service* (Washington, DC), Jan 19, 2015.
39. Monmonier, *Bushmanders and Bullwinkles*, 8.
40. For recent examples of unsuccessful challenges to partisan gerrymandering, see Jeff Barker, "Supreme Court Sidesteps Definitive Ruling in Maryland Congressional Redistricting Case," *Baltimore Sun*, June 18, 2018. For examples of quantitative measures of district shape that have yet to impress the courts, see Richard Barnes and Justin Solomon, "Gerrymandering and Compactness: Implementation Flexibility and Abuse," *arXiv preprint, arXiv:1803.02857* (2018) (accessed October 25, 2018); and Eion Blanchard and Kevin Knudson, "Measuring Congressional District Meandering," *arXiv preprint, arXiv:1805.08208* (2018) (accessed October 25, 2018).
41. Jeff Guo, "Welcome to America's Most Gerrymandered District," *New Republic*, November 8, 2012, https://newrepublic.com/article/109938/marylands-3rd-district-americas-most-gerrymandered-congressional-district (accessed October 25, 2018).
42. Christopher Ingraham, "America's Most Gerrymandered Congressional Districts," *Washington Post*, May 15, 2014.

INDEX

A

B

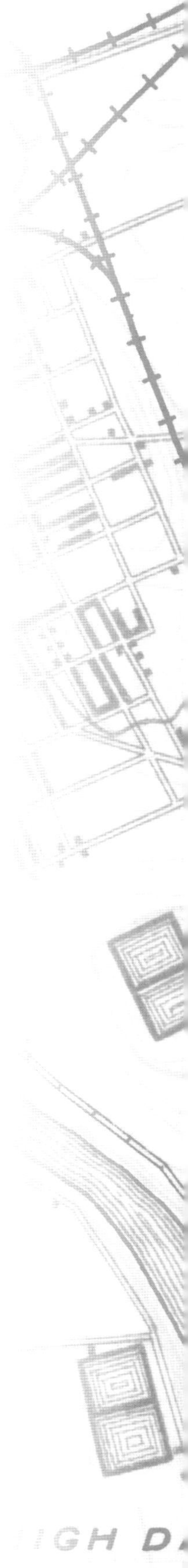

C

D

E

F

G

H

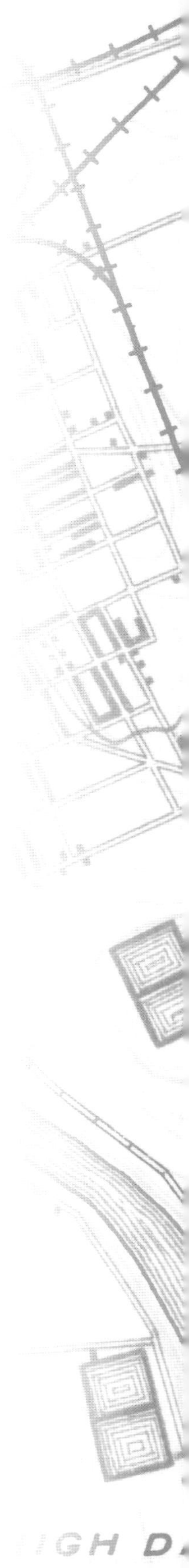

I

K

L

M

N

O

P

Q

R

S

T

U

V

W

Z

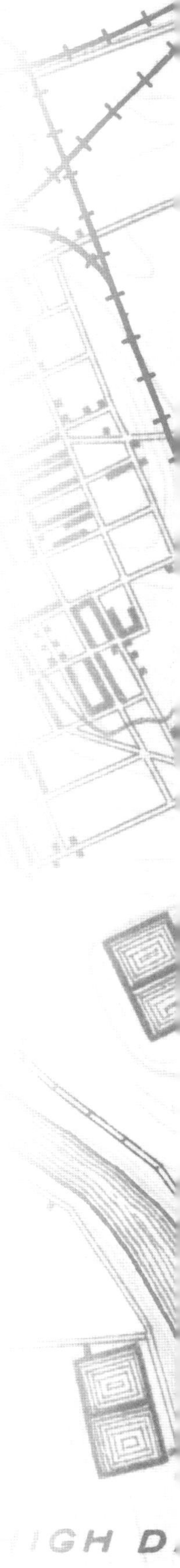